Genetic Codes of
The Artificially-Intelligent Multiverse

by
Vladimir B. Ginzburg

Cover by Eugene B. Ginzburg

Helicola Press
Division of IRMC, Inc.
612 Driftwood Drive
Pittsburgh, Pennsylvania, 15238

Genetic Codes of
The Artificially-Intelligent Multiverse

Published by
Helicola Press
Division of IRMC, Inc.
612 Driftwood Drive
Pittsburgh, Pennsylvania
USA

Printed in the United States of America

ISBN: 978-1-7331402-2-5 (Perfect bound)
ISBN: 978-1-7331402-3-2 (Hardcover)

Current printing (last digit)
10 9 8 7 6 5 4 3 2 1

Contents

My special thanks go to the members of my family: to my still young, gorgeous, friendly and energetic grandson Alex, to my beautiful and talented grandson Asher; to my daughter Ellen with whom I brainstormed some ideas related to my theory and who proofread Summary of this book; to my son Gene whose comments about a multi-level universe made in December of 1992 had triggered my interest in the development of the idea described in this book, and who also designed covers of all my books and helped me to produce them; to my brother Paul who patiently followed my research in this field and provided me with his valuable comments and corrections, and, finally, to my wife Tanya, whose advice and assistance in editing my books were invaluable. Many great memories about my wonderful late parents and very wise grandparents provided me with a needed moral support in writing my books.

ACKNOWLEDGEMENTS

The concept of spiral spacetime is not novel. Introduced at the dawn of science almost 2600 years ago in the ancient Greece as a *vortex* responsible for the formation of the Universe, it had passed through several phases of gaining its strength by absorbing the discoveries made in classical mechanics, hydrodynamics, electromagnetism, atomic & subatomic physics and astrophysics. A list below shows names of philosophers, scientists and mathematicians who either directly or indirectly contributed to the idea of spiral spacetime described in this book.

Ancient Greek Civilization

- Greek natural philosopher Anaximander of Miletus (611-547 BC)
- Greek mathematician and astronomer Pythagoras of Samos (c.560-480 BC)
- Greek natural philosopher Anaxagoras of Glazomenae (500-c.428 BC)
- Greek natural philosopher Empedocles of Acragas (c.490-c.430 BC)
- Greek natural philosopher Leucippus of Miletus or Abdera (5th cent. BC)
- Greek natural philosopher Democritus of Abdera (c.470-c.400 BC)
- Greek mathematician and physicist Archimedes of Syracuse (c.287-212 BC).

Classical Mechanics and Hydrodynamics

- German astronomer and physicist Johannes Kepler (1571-1630)
- French philosopher and mathematician Renè Descartes (1596-1650)
- Dutch physicist and astronomer Christiaan Huygens (1629-1695)
- English physicist and mathematician Isaac Newton (1642-1727)
- German mathematician Gottfried Leibniz (1646-1716)
- French-English mathematician Abraham de Moivre (1667-1754)
- Swiss mathematician Daniel Bernoulli (1700-1782)
- German physicist and physiologist Hermann Helmholtz (1821-1894)
- Italian mathematician Eugenio Beltrami (1835-1899)
- Swedish meteorologist Tor Bergeron (1891-1977)
- German meteorologist Walter Findeisen (1909-1945).

Electromagnetism

- Swedish physicist Emanuel Swedenborg (1688-1772)
- American experimenter and theorist Benjamin Franklin (1706-1790)
- French physicist Charles Coulomb (1736-1806)
- French physicist and mathematician André-Marie Ampère (1775-1836)
- German mathematician Karl Gauss (1777-1855)
- French physicist Augustin Fresnel (1788-1827)
- British chemist Michael Faraday (1792-1867)
- British physicist William Thomson, also known as Lord Kelvin, (1824-1907)
- Scottish physicist Peter Tait (1831-1901)
- British physicist James Maxwell (1831-1879).

Atomic & Subatomic Physics and Astrophysics

- Russian chemist Dmitri Mendeleyev (1834-1907)
- Swedish spectroscopist Johannes Rydberg (1854-1919)
- German physicist Max Planck (1858-1947)
- German physical chemist Walther Nernst (1864-1941)
- German physicist Arnold Sommerfeld (1868-1951)
- New-Zealand-British physicist Ernest Rutherford (1871-1937)
- American amateur scientist, artist and sculptor Walter Russell (1871-1963)
- German mathematician Theodor Kaluza (1873-1916)
- German astronomer Karl Schwarzschild (1873-1916)
- German-Swiss-American physicist Albert Einstein (1879-1955)
- American amateur scientist Victor Schauberger (1885-1958)
- Dutch physicist Neils Bohr (1885-1962)
- Austrian physicist Erwin Schrödinger (1887-1961)
- Dutch physicist Adriaan Fokker (1887-1972)
- British physicist Henry Moseley (1887-1915)
- British chemist Alfred Parson (1889-1970)
- French physicist Louis de Broglie (1892-1987)
- Swedish physicist Oskar Klein (1894-1977)
- Austrian-Swiss-American physicist Wolfgang Pauli (1900-1958),
- German physicist Werner Heisenberg (1901-1976)
- British physicist Paul Dirac (1902-1984)
- Swedish scientist Hannes Alfvén (1908-1995)
- Dutch physicist Hendrik Casimir (1909-2000)
- American theoretical physicist John Wheeler (1911-2008)
- American physicist Winston Bostick (1916-1991)
- American physicist Richard Feynman (1918-1988)
- American physicist Murray Gell-Mann (b. 1929)
- British physicist Roger Penrose (b. 1931)
- American physicist George Zweig (b. 1937)
- American physicist Richard Gauthier (b. 1946).

Thanks to recommendations made by Dr. Akhlesh Lakhtakia (Department of Engineering and Mechanics, Pennsylvania State University), my first three papers describing the earliest versions of my theory were published in *Speculations in Science and Technology* in 1996-1998. I value greatly my two meetings in October 2015 with the Benjamin Powell Professor and Professor of Mathematics, Physics and Business Administration at Duke University Dr. Arlie Oswald Petters who advised me how to introduce my theory to the academic world. I am also thankful to Dr. John Kern II (Department of Mathematics & Computer Science of Duquesne University) for providing me with an opportunity to conduct a seminar related to my theory in their department on February 15, 2017. It was a great opportunity for me to present my theory at the Materials Science & Technology 2017MS&T Conference held in Pittsburgh, PA and at the Materials Science & Technology 2018MS&T Conference held in Columbus OH.

Vladimir B. Ginzburg, Ph.D.

THE ARTIFICIALLY-INTELLIGENT MULTIVERSE (AIM) THEORY IN - BRIEF

Since ancient times some of our inquisitive predecessors used to ask four questions:

1. How was **something** created out of **nothing**?
2. What are the structures and properties of **something**?
3. What kind of laws govern existence of **something**?
4. What is the **purpose** of human beings?

The latest theories of physics and astrophysics proposed during last 120 years shown below only partially answered to the first three questions, leaving the most important fourth question to be speculated by philosophers, historians, religious scholars, writers, artists, archeologists, social workers, futurists, and many other people of various occupations and skills.

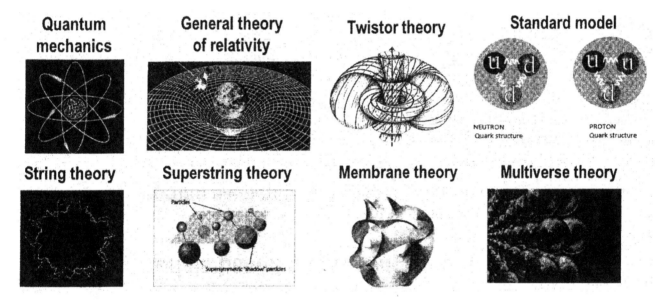

A brief description of the AIM theory is presented in this Introduction in two parts. Part A describes mathematical aspects of the AIM theory. Part B describes the Artificial Intelligence aspects of that theory and the purpose of human beings.

Part A
Mathematical Aspects of the AIM Theory

A1. Selection of prime elements of Nature - Unlike of spacetimes selected by several known theories that have either little or no relations with a real world, the AIM theory selects spacetimes that are two particular cases of a multi-level spiral spacetime called **celestial helicola**. Let us start from describing the simplest 4D spiral spacetime shown in Fig. A1.1. It is made up of two strings, a **leading string** propagating along a straight line with the translational velocity V_t and the spiral **trailing string** propagating synchronously with the leading string with the spiral velocity V_s. The **spiral velocity** has two components: the **translational velocity** V_t and the **rotational velocity** V_r. In this figure r is the radius, λ is the **wavelength** and φ is the **steepness angle** of trailing string.

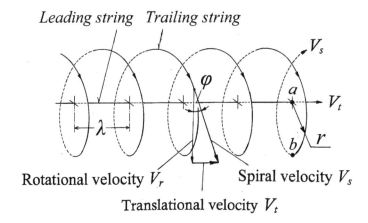

Leading string   Trailing string

Rotational velocity V_r Spiral velocity V_s

Translational velocity V_t

Figure A1.1. The simplest 4D spiral spacetime.

Do not be surprised if you never heard about the celestial helicola. Most of us are familiar with a known illustration of a solar planetary system shown in Fig. A1.2 (top) in which a stationary Sun is orbited by planets moving along elliptical paths around the Sun. We know nowadays that as the Earth moves around the Sun, the Sun also moves around the center of Milky Way. Consequently, the Earth moves around the Sun along a spiral path as shown in Fig. A1.2 (bottom).

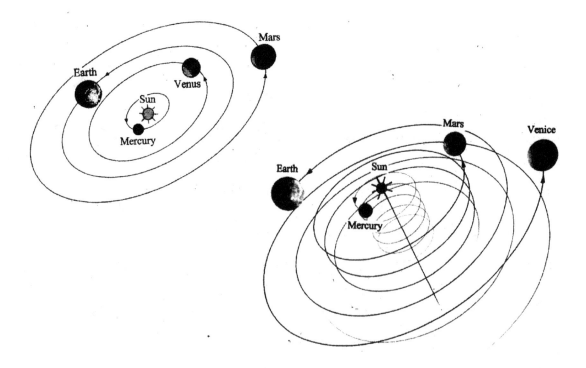

Figure A1.2. Outdated (top) and updated (bottom) presentations of our solar planetary system.

There is a leading string and a trailing string at each level of celestial helicola. A trailing string of each previous level of helicola becomes a leading string of the next level.

A2. Spacetime structures of toryces and helyces –

A2. Spacetime structures of toryces and helyces – The spacetime structures of the 4D spiral spacetime elements of the Multiverse *toryx* (Fig. A2.1) and *helyx* (Fig. A2.2) are based on the same principle as the celestial helicola.

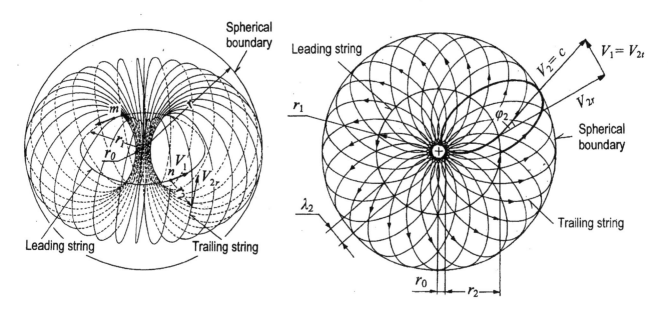

Figure A2.1. Toryx basic spacetime structure: isometric (left) and top view (right).

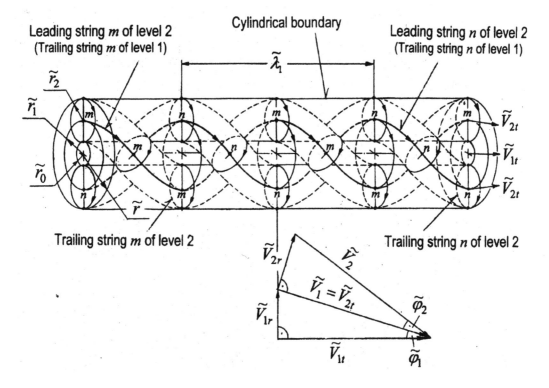

Figure A2.2. Structure of the first and second helyx levels.

A3. Spacetime covered by the AIM theory – Currently, the AIM theory covers only *our Multiverse* existing in six quantum spacetime levels ranging from *L0* to *L5* as shown in Fig. A3. Future versions of the AIM theory may describe the other Multiverses residing inside the *Infiverse*.

There are two kinds of toryces in each quantum spacetime level of our Multiverse: micro-toryces and macro-toryces shown respectively at the left and right sides of Fig. A3. What we call "our Universe" belongs to the spacetime level *L1*.

The micro-toryces form elementary matter particles that consequently form nucleons, atoms and *celestial bodies*. The macro-toryces form *celestial fields* associated with celestial bodies. These fields are responsible for interactions between celestial bodies.

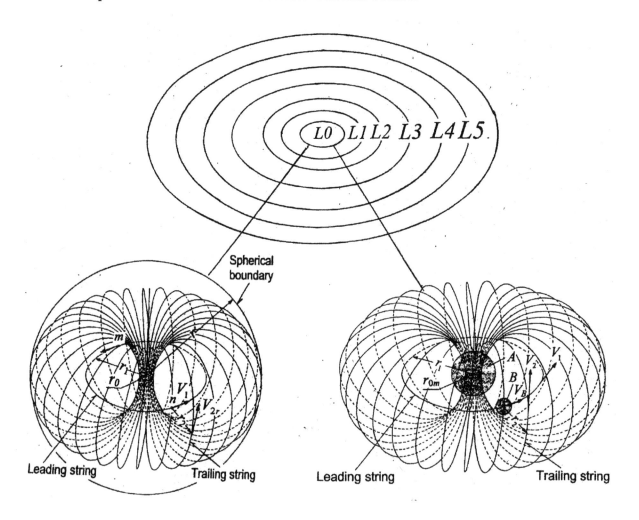

Figure A3. Quantum spacetime levels of our Multiverse and its components residing inside quantum vacuum infinity range.

A4. Creation of toryces and helyces of the micro-world - Toryces of the microworld can be of two kinds: *virtual toryces* and *4D spacetime toryces*. The virtual toryces appear spontaneously inside *quantum vacuum* expanding from unity (± 1) towards both *infinity* $(\pm\infty)$ and *infinility* $(\pm 0 = 1/\pm\infty)$ as shown in Fig. A4, and they have unrestricted numbers of degrees of spacetime freedom.

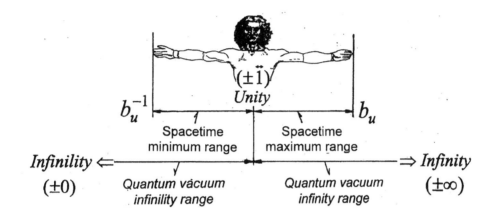

Figure A4. The ranges occupied by virtual and spacetime toryces.

The 4D spiral spacetime toryces are created from rare virtual toryces having limited numbers of degrees of freedom defined by the *toryx genetic codes* making them to appear in the forms shown in Fig. A2.1. As shown in Fig. A4, the 4D spacetime toryces reside within the spacetime ranges of relative radii of their spherical boundaries expanding from unity (± 1) towards both the ultimate maximum relative radius b_u and the minimum relative radius $b_u^{-1} = 1/b_u$. The 4D spacetime toryces exist at certain excitation, oscillation and spacetime quantum states. They sustain their existence by cyclic absorption and release of spacetime.

The helyces of the micro-world (Fig. A2.2) are emitted when excitation, oscillation and spacetime quantum states of their parental toryces are reduced from higher to lower quantum states. The helyx degrees of spacetime freedom are limited by the *helyx genetic codes*.

A5. Topological inversions of toryces and helyces – Based on their genetic codes, both toryces and helyces are polarized by inversions of their strings inside out allowing them to exist in four kinds of topologically-polarized states. Figure A5 (top) shows that as the radius of the toryx leading string decreases from positive to negative infinity, the steepness angle of toryx trailing string φ_2 increases from 0^0 to 360^0. Consequently, four kinds of topologically-polarized toryces are formed:

- Real negative toryces - *Trailing string* becomes inverted at $\varphi_2 = 90^0$
- Real positive toryces - *Wavelength of trailing string* becomes inverted at $\varphi_2 = 180^0$
- Imaginary positive toryces - *Leading string* becomes inverted at $\varphi_2 = 270^0$
- Imaginary negative toryces - *Entire toryx* becoming inverted at $\varphi_2 = 0^0/360^0$.

Four topologically-polarized helyces are formed in a similar way – see Fig A5 (bottom).

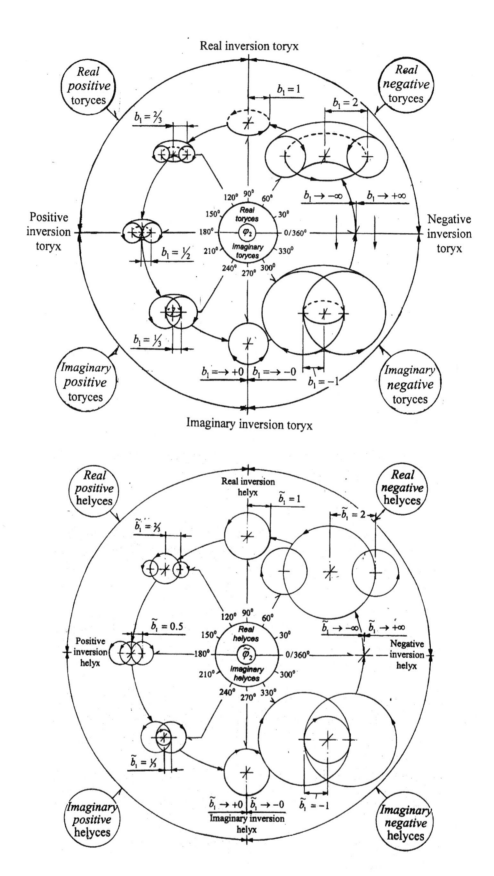

Figure A5. Topological transformations of toryces (top) and helyces (bottom).

A6. Formation of elementary particles –

Elementary matter and radiation particles are formed by unification of respectively polarized toryces and helyces of the micro-world. Elementary matter and radiation particles sustain their existence by periodic absorption and release of spacetimes. The elementary matter particles are divided into two groups, *mutually-sustained particles* and *self-sustained particles*.

   Mutually-sustained particles are formed by the unification of topologically-polarized toryces. Figure A6 shows a formation of four mutually-sustained elementary matter particles: *electrons*, *positrons*, *ethertrons* and *singulatrons*. In these particles, the translational velocities of trailing strings of their real toryces are below the velocity of light during each cycle, while in their imaginary toryces the translational velocities exceed the velocity of light. Consequently, the imaginary toryces absorb spacetime, while the real toryces release it.

   Self-sustained particles are able to sustain their existence by themselves, because the translational velocities of trailing strings of their constituent toryces vary cyclically below and above velocity of light during each cycle.

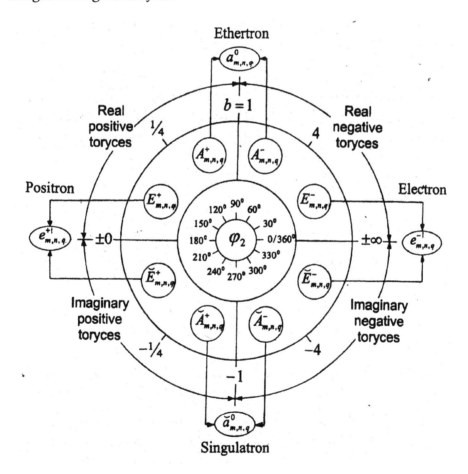

Figure A6. Four elementary matter particles formed by mutually-sustained toryces.

   Topologically-polarized helyces are formed when their parental topologically-polarized toryces are transferred from higher to lower excitation and oscillation quantum states. The elementary radiation particles are formed by the unification of topologically-polarized helyces.

A7. Symmetrical polarization of toryces and helyces – Figure A7 uses dotted horizontal and vertical lines to show symmetrical polarizations between the vorticities V of toryx (top) and between the vortices $\widetilde{V}$ of helyx (bottom).

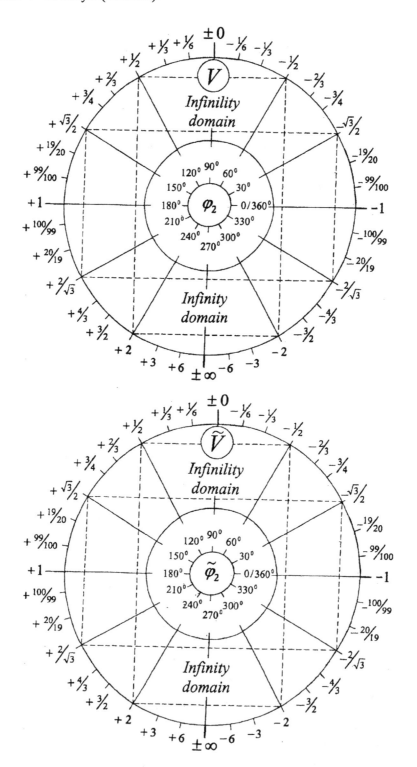

Figure A7. Symmetrically-polarized vorticities of toryx (top) and helyx (bottom).

A8. Excitation and oscillation of toryces

A8. Excitation and oscillation of toryces – Based on their genetic codes, the toryces may exist in excitation and oscillation quantum states (Fig. A8). During excitation of a toryx the radius of toryx leading string r_1 increases, while its eye radius r_0 remains constant. During oscillation of a toryx, its radius of leading string r_1 and its eye radius r_0 change with the same rate.

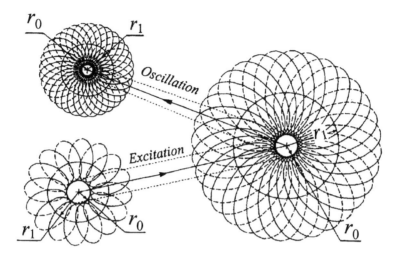

Figure A8. Excitation and oscillation of toryces.

A9. Generic properties of toryces and helyces

A9. Generic properties of toryces and helyces - Unlike other proposed candidates for prime elements of a real world, toryx and helyx possess ***generic properties*** that can be found in many entities of our Universe. Take those properties away and our Universe, as we know it, will cease to exist.

Motion – Both leading and trailing strings of toryces and helyces are in a state of constant motion. All entities in our Universe are in the motion state.

Spirality – Trailing strings of toryces and helyces propagate along spiral paths. Each celestial body spirals around another celestial body that, in its turn, spirals around another celestial body, etc.

Propagation with velocity of light – Trailing strings of toryces and helyces propagate along their spiral paths with velocity of light. All electromagnetic waves in our Universe propagate in vacuum with velocity of light.

Limitation of spacetime freedom – Toryces and helyces have limited degrees of spacetime freedom defined by their genetic codes. The spacetime freedom is limited in all entities of our Universe, except for quantum vacuum.

Planetary motion – Leading strings of toryces follow a law of planetary motion for which the Kepler's third law of planetary motion is a particular case.

Symmetry – Symmetry exists between several parameters of toryces and helyces. Symmetry is well known property of many entities of the Universe.

Polarization – Toryces and helyces exist in topologically-polarized states. Atoms are made up of electrically-polarized electrons and nuclei, while many complex elements are made up of chemically-polarized acids and bases.

Cyclic sustaining of existence – Toryces and helyces sustain their existence by cyclic absorption and release of spacetime. All living entities sustain their existence by cyclic absorption and release of energy.

Stable coexistence – Toryces and helyces coexist stably with their respective topologically-polarized toryces and helyces and form respectively particles of micro- and macro-worlds. Electrons and protons coexist stably in atoms, while acids and bases coexist stably in DNA.

Expansion and contraction – Toryces and helyces are able to expand and contract. This capability can be found in all entities of our Universe.

Quantization – Toryces and helyces exist in quantum excitation states. The atomic electrons also exist in quantum excitation states.

Emission of radiation – Excited toryces are capable of emitting the radiation in the form of helyces. Excited atomic electrons emit electromagnetic radiation in the form of photons.

A10. Formation of nucleons – Nucleons (protons and neutrons) are made up of three parts: *nucleon crystal*, *nucleon core* and *nucleon leptons*.

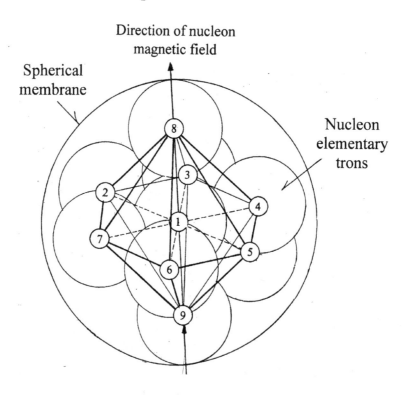

Figure A10.1. Isometric view of a nucleon crystal.

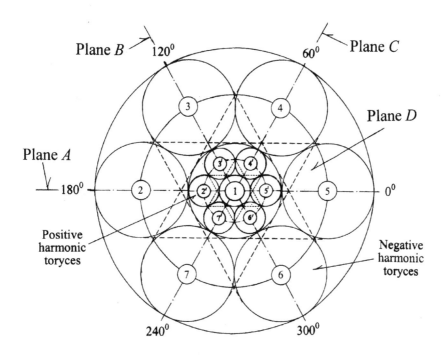

Figure 10.2. Cross-section of outer and inner parts of the nucleon crystal

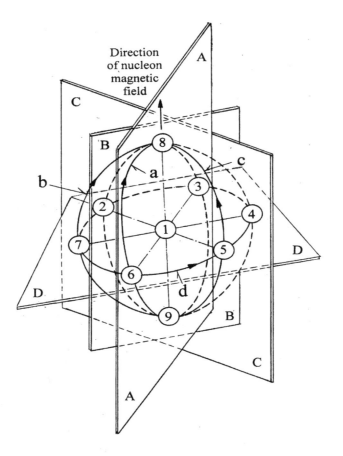

Figure 10.3. Isometric view of either outer or inner part of the nucleon crystal.

The main purpose of nucleon crystals (Fig. 10.1) is to retain excited ethertrons and singulatrons, the constituents of nucleon cores. This is accomplished by locating the ethertrons and singulatrons inside the cavities formed around the center 1 and the vertices 2 through 9 of a nucleon bi-pyramid hexagonal crystal (Fig. 10.2) made up of outer and inner parts. The inner part is located inside the outer part and it is three times smaller than the outer part. Each part has the center 1 and eight vertices 2 through 9. As shown in Figs. 10.2 and 10.3, the vertices of the outer and inner parts are respectively located at the intersections of leading string strings a, b, c and d of negative and positive toryces. Figures 10.4 and 10.5 show nucleon crystal structures of deuterium and tritium.

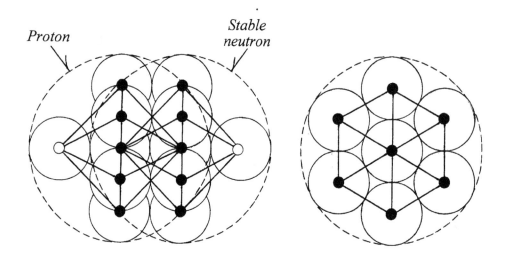

Figure 10.4. Structure of the deuterium nucleus:
front view (left) and side view (right).

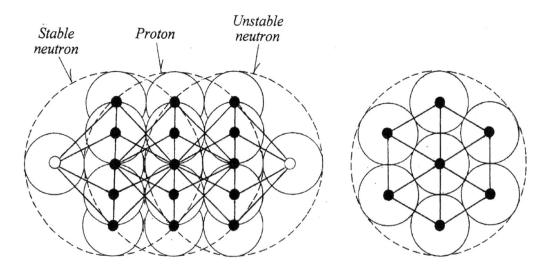

Figure 10.5. Structure of the tritium nucleus:
front view (left) and side view (right).

A11. Extrapolation Properties of Toryces and Helyces – It means that, besides the original applications of their genetic codes for defining structures and properties of entities of the level *L1* of the Multiverse (that is our Universe), those genetic codes are also applicable for defining structures and properties of entities of other quantum spacetime levels of the Multiverse.

A12. Effects of spacetime levels of the Multiverse – As shown in Fig. A12, the entities of all levels of the Multiverse are made up mostly out the same elements as the entities of the spacetime level *L1*, but with several important differences. As the spacetime level of the Multiverse increases, the properties of their entities change substantially as described below:

- The orbital radii of atomic electrons significantly increase.
- Both mass and strength of atomic nucleons increase and they contain much more energy.
- Materials of the levels of the Multiverse higher than the level *L1* emit and absorb rays propagating progressively much faster than velocity of light.

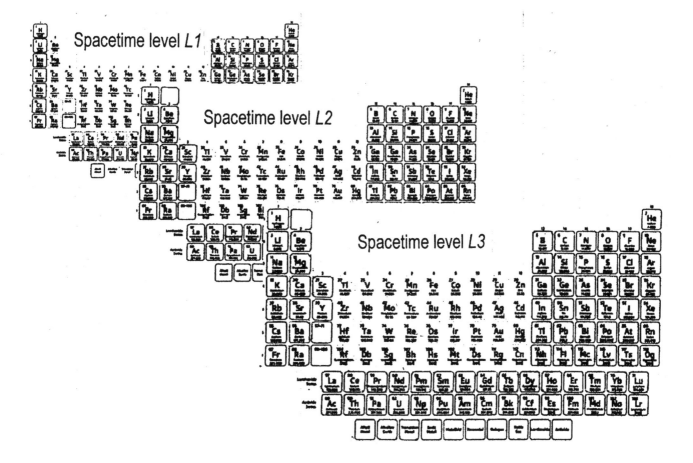

Figure A12. Tables of elements of three spacetime levels of the Multiverse.

A13. Outverted, inverted and imaginary macro-toryces – In the macro-world, three kinds of stars can be identified: *outverted*, *inverted* and *imaginary* as shown in Fig. A13. This classification is based on the relationship between the outer radius r_b of a star and the eye radius r_{jb} of the macro-toryces forming the spacetime fields associated with stars. Our Sun, neutron stars and pulsars are examples of outverted stars, while the black holes of galaxies are examples of inverted stars, and the black holes of quasars are examples of imaginary stars.

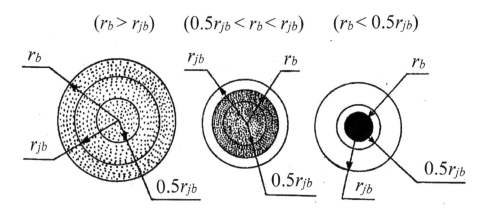

$$(r_b > r_{jb}) \qquad (0.5r_{jb} < r_b < r_{jb}) \qquad (r_b < 0.5r_{jb})$$

Figure A13. Three kinds of stars: outverted (left), inverted (center) and imaginary (right).

A14. Main features of math of the AIM theory – The math of the AIM theory has three main features:

Unified equations – The math of the AIM theory employs mostly the same equations to describe structures and properties of entities of both micro- and macro-worlds residing in all spacetime levels of the Multiverse.

Universal spacetime units – What we call matter, field, charge, magnetism, gravity, etc. are merely manifestations of the 4D spiral spacetimes that are at the core of the Multiverse. The math of the AIM theory expresses all physical properties, such as masses, charges, magnetic fields, etc. in the universal spacetime units. It also provides a "Rosetta Stone" establishing a correlation between physical units invented by human beings and spacetime units understandable by civilized being of the Multiverse.

Significantly reduced complexity – The math of the AIM theory is mainly based on 3-dimensional elementary math that is modified to be compatible with the 4-dimensional spacetime. Consequently, because of its simplicity, this theory can be successfully taught in high schools.

A15. Confirmations & predictions of the AIM theory – This theory confirms numerous experimental data related to the properties of entities of our Universe. It also predicts still unknown structures and properties of entities of both micro- and macro-worlds. Based on the discovery of unique properties of entities of various spacetime levels of the Multiverse, the AIM theory made its most important prediction related to the unified purpose of human beings.

A16. Spacetime Genesis of the Multiverse – Below are ten main conclusions of the AIM theory related to that subject.

1. What we call matter, field, charge, magnetism, gravity, etc. are merely manifestations of the 4D spiral spacetimes that are at the core of the Multiverse.
2. The 4D spiral spacetime entities toryces can be created out of virtual (incomplete) toryces of quantum vacuum and serve as the prime elements of elementary matter particles.

3. The toryces may exist in specific excitation and oscillation quantum states.

4. The 4D spiral spacetime entities helyces are emitted by their parental excited and oscillated toryces and serve as the prime elements of elementary radiation particles.

5. The spacetime properties of both toryces and helyces are governed by their respective genetic codes that limit their respective degrees of freedom.

6. Elementary matter particles are formed my unification of polarized toryces, while elementary radiation particles are formed by unification of polarized helyces.

7. Both toryces and helyces sustain their existence by periodic absorption and release of spacetime.

8. Celestial bodies are assemblies of elementary matter particles.

9. Associated with each celestial body are the macro-toryces, the toroidal spiral fields responsible for the interaction between celestial bodies.

10. The Multiverse may exist in several quantum spacetime levels.

Based on the discovered capabilities of the prime elements of elementary particles, the AIM theory proposes that there is the *Supreme Law of the Multiverse* that states:

The Multiverse and all its entities exist, because of their abilities to be created and to sustain their existence.

Part B
The Artificial-Intelligence Aspects of the AIM Theory

The artificial-intelligence aspects of the AIM theory are based on the branch of science called *cybernetics*.

B1. Basic principles of cybernetics – That branch of science was proposed by the American engineer Norbert Wiener in 1948. Figure B1 shows a block diagram of its basic system called *the closed-loop feedback control system*. It has the following main components:

- Sensors (eyes, nose, skin)
- Actuators (mouth, teeth, hands, legs and wings for birds)
- Comparator and controller (parts of brains)
- Communication lines between the above components, and
- Entities responsible for absorption, digestion, distribution and release of energy.

Norbert Wiener discovered that cybernetics is at the core of existence of both animals and human beings. For instance, when an early human being wanted to warm up himself by starting a camp fire, his skin performed like a temperature sensor providing a measured signal to his brain. The brain comparator then compared the measured signal with a reference value of temperature at which that human being felt comfortable. If there was a difference between those two values, then the comparator sent an error signal to the brain controller that decided how much wood has to be used in the camp fire. Consequently, the brain controller sent a command signal to the human being's hands (actuators) that followed that command until the temperature felt by human being's skin became sufficiently close to the skin temperature reference value.

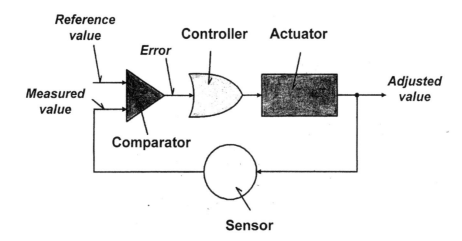

Figure B1. Closed-loop feedback control system.

The process of creation of contemporary human beings can be divided into four main ages.

1. The age of formation of the earth incubator – That age begun 4.57 billion years ago after creation of our solar system followed by a violent period of the Earth during which it was bombarded by numerous celestial bodies and devastated by underground volcanic forces. During that time the goldilocks Earth was gradually transformed into an ***incubator*** made up of dry lands, oceans, lakes, rivers, atmosphere, magnetic field, trees, plants and minerals providing the upcoming human beings and other living creatures with environmental conditions, energy, air, water and food resources necessary for their creation and survival.

2. The earth exploration & industrial revolution age – First appearing in Africa about 315 thousand years ago, the human beings spread around the Earth. The basic principles of cybernetics applied to the human beings and other living creatures in the same way, but the human being had something in their DNAs that made them capable of making the components of the closed-loop feedback controls systems much more efficient than was originally created by Nature. That feature of the human beings was most visibly revealed during industrial revolution when the human beings started to develop machines, cars, trucks, trains, ships, submarines and airplanes.

3. The computer age - Although the basic principles of cybernetics were still used in many control systems, by 1956, the computers had begun to perform much more sophisticated functions encouraging the scientists to use the name "***artificial intelligence***", or shortly AI, instead of cybernetics. The implementation of the AI helped human beings to increase further their physical strength and productivity, making them superior to all other living creatures on the planet Earth.

4. The current cosmos exploration age - This is a current age during which the human beings begun to use AI in hostile environment of outer space. It started in October of 1957 when the Soviet Union launched the first satellite into space. Nowadays more than a thousand of satellites and two space stations spin around the Earth fulfilling various functions vital for survival of human beings. Beginning from 1969, several astronauts visited the Moon and several drones roamed on the surfaces of the Moon and Mars. Scientists and engineers no longer wait passively for some dangerous asteroids to hit our planet, but design systems capable of tracking their paths and develop the ways to avoid our collisions with them. They are also no longer satisfied with a current supply of minerals available on our planet and getting ready for mining the Moon, Mars and other celestial bodies of our solar system to discover new minerals with extraordinary properties.

B2. Creation of the AI Multiverse - The Artificially-Intelligent Multiverse (AIM) theory envisions the following sequence of development of the Artificially-Intelligent Multiverse.

The human beings, existing in the first spacetime level $L1$ of the Multiverse, will create the first level Artificial Intelligence (AI1) allowing them to explore celestial bodies located in our solar system and to discover materials that belong to the second spacetime level $L2$ of the Multiverse. Those materials will be then used to create the second level Artificial Intelligence (AI2). Consequently, the materials discovered with a help of the AI2 will be used for creation of the third level Artificial Intelligence (AI3) that belong to the third spacetime level $L3$ of the Multiverse. This process will eventually lead to the creation of the fourth and fifth level Artificial Intelligences (AI4) (AI5) that respectively belong to the fourth and fifth spacetime levels $L4$ and $L5$ of the Multiverse.

A transfer to a next level of the Multiverse would require to input a significant amount of energy comparable with the amount of energy released during collisions between massive celestial bodies. Because the planet Earth is covered by thick layers of atmosphere and ocean water, the impact of its collisions with asteroids is significantly reduced, explaining why we so far were not able to find any entities that belong to the spacetime levels greater than $L1$.

The AIM theory, however, predicts that some particles discovered during collisions of protons at the CERN proton-proton collider, like Z and Higgs bosons, might belong to the spacetime level $L2$ of the Multiverse. The AIM theory also predicts that those particles will travel with velocities that are significantly faster than velocity of light, making their detections very difficult. Those material are expected to be created during very strong collisions between celestial bodies. In our solar system, Mars will most likely be a prime candidate, followed by our Moon and possibly by the Earth after its collision with an asteroid 65 million years ago that killed dinosaurs.

B3. Maintaining required rate of development of the AI Multiverse - Currently, celestial bodies residing in all levels of the Multiverse are not equipped with closed-loop feedback control systems allowing them to avoid collisions with one another. However, as the Multiverse becomes overcrowded, a probability of occurrences of those collisions increases. Therefore, to comply with the Supreme Law of the Multiverse the development of the AI Multiverse must be fast enough to avoid its self-destruction. The AIM theory proposes that the Multiverse employs three kinds of stimulating mechanisms to assure a compliance of the developers of the AI Multiverse with the Supreme Law of the Multiverse: *competition, wars* and *pandemics*.

Competition - The biologically built-in desire of human beings to compete with one another stimulates them to develop more advanced technologies much faster. As long as this rate is adequate for survival of the Multiverse, the coexistence between people remains relatively peaceful. History tells us, however, that those periods of time were very short.

Wars - Conversely, when the rate of development of the AI Multiverse becomes too slow for the survival of the Multiverse, the level of polarization between people increases and ends up with mutually-destructive wars. The history of the human beings confirms that the wars stimulate people to become more innovative and eventually helped them to start exploration of outer space shortly after the most destructive World War II.

Pandemics – Those unfortunate events usually come when many human beings become reluctant to use the already developed advanced technologies. Many of us had recently witnessed how the COVID-19 pandemic forced many people to start replacing very expensive air travels and inefficient educational, commercial and communication systems with much more advanced systems utilizing online virtual technologies, including video conferencing, work from home and online shopping.

B4. The unified purpose of human beings - It took Nature billions of years to create an *incubator* called the planet Earth providing the upcoming human beings with environmental conditions and necessary energy, mineral and food resources for their survival.

Human beings became the most advanced creature on the planet Earth who initially supplemented their biological capabilities with much stronger artificial tools and currently invented and began to use amazing capabilities of the artificial intelligence capable of surviving and operating successfully in the outer space beyond the planet Earth.

The development of the artificial intelligence, however, had now reached a level at which the human beings are able to destroy themselves by using nuclear weapons, while being occupied with arguments about much less important issues. Here is why the AIM theory may become helpful for discovering the unified purpose of human beings:

- The AIM theory shows that our Universe is not alone, but belongs to the spacetime level $L1$ of the Multiverse existing in several other spacetime levels. The spacetime level $L1$ was probably the best for creation of human beings inside the incubator formed by the goldilocks planet Earth and also for a start of creation of the first non-human AI by human beings.

- There are three main deficiencies of the spacetime level $L1$ making it practically very difficult to use for exploration of outer space beyond our Moon:

  (1) Materials available in that level are made up of very small atoms that are heavy and not strong,
  (2) Level of energy available on the Earth is too low, and
  (3) Transmission of information is not possible with velocities faster than the velocity of light.

- Fortunately, as the spacetime levels of the Multiverse increase beyond the spacetime level $L1$, the atoms become significantly larger, lighter and stronger, the available level of energy also significantly increases, and the information can be transmitted with velocities that are progressively much faster than velocity of light.

The time had finally arrived for the human beings to learn about their unified purpose:

> ## The unified purpose of human beings is to create
> ## The Artificially-Intelligent Multiverse.

B5. Possible applications of the AIM Theory - Some predictions of the AIM theory may stimulate discoveries and developments in the following areas of knowledge.

Mathematics – Development of new branches of mathematics related to the 4D spiral spacetime topology, trigonometry and number theory.

Particle physics – Novel explanation of structures and properties of elementary matter and radiation particles, nucleons, atoms, and thin films.

Astrophysics –Novel explanation of structures and properties of neutron stars, pulsars, black holes of galaxies and quasars.

Material science & technology – Development of materials with considerably greater ranges of their properties, including the materials made up of elements existing in the higher spacetime levels of the Multiverse.

Communication science & technology – Development of communication systems capable of transmitting the information with velocities significantly exceeding the velocity of light.

Computer science & technology – Development of a more compact computer technology with increased speed and more precise accuracy of calculations.

Artificial Intelligence science & technology – Development of artificial intelligence that belong to the higher spacetime levels of the Multiverse.

Energy science – Development of much more efficient methods of generation and storage of energy with minimal detrimental impact on environment.

Virtual sciences & technology – Development of more advanced scientific programs and technologies to be used for more efficient virtual education, conferences, work out of office, entertainment, and travels.

History of competition, wars and pandemics – Discovery of effects of competition, wars and pandemics on the development of science and technology.

Earth system science – Development of effective methods of reducing devastating destructions and losses of human lives caused by tornadoes, hurricanes, floods, fires, earthquakes, volcanoes, super-volcanoes and both little and big ice ages.

Biology, chemistry, medicine and genetic engineering – Development of effective methods to improve health, longevity of human beings, and to increase their creative capabilities.

Music, art & graphic design - Development of products of music, art and graphic design resonating with the spacetime properties of the Multiverse.

People of various fields of knowledge - Development of information helping the human beings to understand their long-term unified purpose.

<u>Notes:</u>

PART 1

Abstract Mathematics of a Toryx

1. TORYX BASIC STRUCTURE & PARAMETERS

1.1 Toryx Basic Structure

Toryx basic structure consists of a double-circular *leading string* and a double-toroidal *trailing string*, both residing inside a *spherical boundary* as shown in Figs. 1.1.1 – 1.1.3.

Toryx leading string – The toryx double-circular leading string appears like two circular lines with the radius r_1 formed by the moving points m and n shown in Figure 1.1.1. It can be thought as a particular case of a double-helical spiral in which the translational velocity V_{1t} is equal to zero, so the rotational velocity V_{1r} is equal to the spiral velocity V_1. Thus,

$$V_{1t} = 0 \qquad\qquad (1.1\text{-}1)$$

$$V_{1r} = V_1 \qquad\qquad (1.1\text{-}2)$$

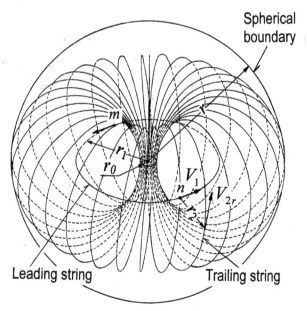

Figure 1.1.1. Isometric view of a toryx.

Toryx trailing string – The toryx double-toroidal trailing string has the radius r_2 and a circular opening at the toryx center with the radius r_0 called the **toryx eye**. It can be thought as a dynamic double-toroidal spiral line in which each branch propagates along its toroidal spiral path with the spiral velocity V_2 that has two components, the translational velocity V_{2t} and the rotational velocity V_{2r}, with all three velocities related to each other by the Pythagorean Theorem:

$$V_2 = \sqrt{V_{2t}^2 + V_{2r}^2}$$

$$(1.1-3)$$

The trailing string propagates synchronously with the leading string. Therefore, the translational velocity of trailing string V_{2t} is equal to the rotational velocity V_{1r} and the spiral velocity V_1 of leading string as given by:

$$V_{2t} = V_{1r} = V_1$$

$$(1.1-4)$$

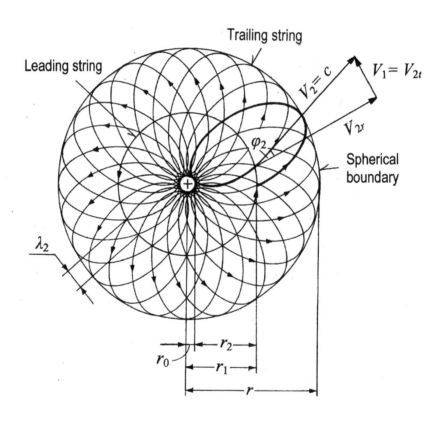

Figure 1.1.2. Top view of a toryx.

In Fig. 1.1.2, φ_2 is the **steepness angle** of toryx trailing string corresponding to the middle point a of toryx trailing string shown in Fig. 1.1.3.

The radius of spherical boundary r is equal to:

$$r = r_1 + r_2 \qquad (1.1\text{-}5)$$

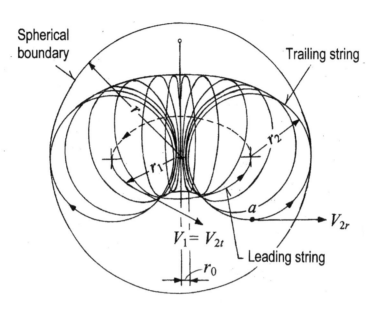

Figure 1.1.3. Cross-section of a toryx.

Toryx spins – Toryx has two spins, leading and trailing, with both of them defined by the right-hand rule as shown in Fig. 1.1.4. The toryx leading and trailing spins depend on the directions of the rotational velocities of toryx leading string V_{1r} and toryx trailing string V_{2r}.

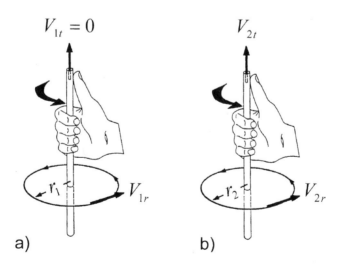

Figure 1.1.4. The toryx leading spin (a) and trailing spin (b) defined by the right-hand rule.

1.2 Toryx Basic Spacetime Parameters in Absolute Units

The values of some toryx basic spacetime parameters are dependent on the position of a point of measurement of these parameters along the toryx trailing string. In the nomenclature of toryx parameters shown below, the symbols of the position-dependent parameters marked with star (*) correspond to the middle point a of toryx trailing string shown in Fig. 1.1.3.

f_0 = toryx base frequency

f_1 = frequency of toryx leading string

f_2 = frequency of toryx trailing string

L_1 = spiral length of one winding of toryx leading string

L_2 = spiral length of one winding of toryx trailing string

L_0 = circular length of toryx eye

r = radius of toryx spherical boundary

r_0 = toryx eye radius

r_1 = radius of toryx leading string

r_2 = radius of toryx trailing string

T_1 = period of toryx leading string

T_2 = period of toryx trailing string

V_1 = spiral velocity of toryx leading string

V_{1t} = translational velocity of toryx leading string

V_{1r} = rotational velocity of toryx leading string

V_2 = spiral velocity of toryx trailing string

V_{2t} = translational velocity of toryx trailing string*

V_{2r} = rotational velocity of toryx trailing string*

w_1 = the number of windings of toryx leading string

w_2 = the number of windings of toryx trailing string

λ_1 = wavelength of toryx leading string

λ_2 = wavelength of toryx trailing string*

φ_1 = steepness angle of toryx leading string

φ_2 = steepness angle of toryx trailing string*.

1.3 Toryx Basic Genetic Codes in Absolute Units

The toryx basic genetic codes include three basic equations limiting the degrees of freedom of several toryx parameters (see Exhibit 1.3). These equations provide the simplest way to derive the relationships between all spacetime parameters of toryx. In spite of their outmost simplicity, these equations provide toryces with amazing spacetime properties, including a capability to exist in four unique topologically-polarized states within the range of the radius of toryx leading strings r_1 extending from negative to positive infinity $(-\infty < r_1 < +\infty)$ as will be described in Chapter 5.

Exhibit 1.3. Toryx basic genetic codes in absolute units.

- The length of one winding of toryx trailing string L_2 is equal to the length of one winding of toryx leading string L_1:

$$L_2 = L_1 = 2\pi r_1 \qquad (1.3\text{-}1)$$

- The difference between the radius of toryx leading string r_1 and the radius of toryx trailing string r_2 is constant and equals to the toryx eye radius r_0:

$$r_0 = r_1 - r_2 = const. \qquad (1.3\text{-}2)$$

- The spiral velocity of toryx trailing string V_2 is constant at each point of its spiral path:

$$V_2 = \sqrt{V_{2t}^2 + V_{2r}^2} = c = const. \qquad (1.3\text{-}3)$$

Figure 1.3 shows the toryx basic spacetime parameters in absolute units corresponding to the middle point a of toryx trailing string (Fig. 1.1.3).

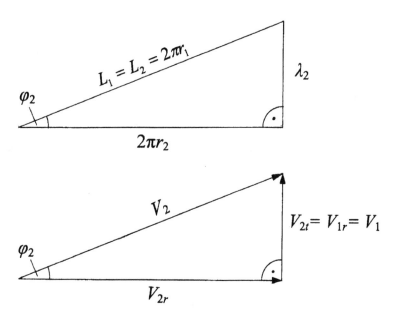

Figure 1.3. Toryx basic spacetime parameters in absolute units corresponding to the middle point a of toryx trailing string (Fig. 1.1.3).

1.4 Toryx Basic Spacetime Parameters in Relative Units

The toryx basic spacetime parameters can be simplified by expressing them in relative units in respect to the constant toryx parameters: the toryx eye radius r_0, the velocity of light c and the toryx base frequency f_0 as shown in Table 1.4.

Table 1.4. Toryx relative spacetime parameters.

Toryx relative parameters	Equations
Radius of toryx spherical boundary	$b = r/r_0$
Radius of toryx leading string	$b_1 = r_1/r_0$
Radius of toryx trailing string	$b_2 = r_2/r_0$
Length of toryx leading string	$l_1 = L_1/2\pi r_0$
Length of toryx trailing string	$l_2 = L_2/2\pi r_0$
Period of one winding of toryx leading string	$t_1 = T_1 f_0$
Period of one winding toryx trailing string	$t_2 = T_2 f_0$
Spiral velocity of toryx leading string	$\beta_1 = V_1/c$
Translational velocity of toryx leading string	$\beta_{1t} = V_{1t}/c$
Rotational velocity of toryx leading string	$\beta_{1r} = V_{1r}/c$
Spiral velocity of toryx trailing string	$\beta_2 = V_2/c$
Translational velocity of toryx trailing string	$\beta_{2t} = V_{2t}/c$
Rotational velocity of toryx trailing string	$\beta_{2r} = V_{2r}/c$
Frequency of toryx leading string	$\delta_1 = f_1/f_0$
Frequency of toryx trailing string	$\delta_2 = f_2/f_0$
Wavelength of toryx leading string	$\eta_1 = \lambda_1/2\pi r_0$
Wavelength of toryx trailing string	$\eta_2 = \lambda_2/2\pi r_0$

The ***toryx base frequency*** f_0 corresponds to the case when $r_1 = r_0$ and it is equal to:

$$f_0 = \frac{c}{2\pi r_0}$$

(1.4-1)

1.5 Toryx Basic Genetic Codes in Relative Units

Exhibit 1.5 shows three toryx basic genetic codes in relative units.

Exhibit 1.5. Toryx basic genetic codes in relative units.

- The relative length of one winding of toryx trailing string l_2 is equal to the relative length of one winding of toryx leading string l_1:

$$l_2 = l_1 \qquad (1.5\text{-}1)$$

- The toryx relative eye radius b_0 is equal to 1:

$$b_0 = b_1 - b_2 = 1 \qquad (1.5\text{-}2)$$

- The relative spiral velocity of toryx trailing string β_2 is equal to 1 at each point of its spiral path:

$$\beta_2 = \sqrt{\beta_{2t}^2 + \beta_{2r}^2} = 1 \qquad (1.5\text{-}3)$$

Figure 1.5 shows the toryx relative spacetime parameters corresponding to the middle point a of toryx trailing string (Fig. 1.1.3).

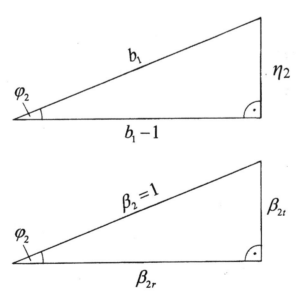

Figure 1.5. Toryx relative spacetime parameters corresponding to the middle point a of toryx trailing string (Fig. 1.1.3).

1.6 Summary of Derived Toryx Equations

Table 1.6.1 provides a summary of derived equations for main relative spacetime parameters of toryx as functions of the relative radius of toryx leading string b_1.

Table 1.6.1. Equations for relative spacetime parameters of toryx leading and trailing strings as functions of the relative radius of toryx leading string b_1.

Relative parameter	Leading string Eq. (a)	Trailing string Eq. (b)
Radius Eq. (1.6-1)	$b_1 = \dfrac{r_1}{r_0}$	$b_2 = \dfrac{r_2}{r_0} = b_1 - 1$
Wavelength Eq. (1.6-2)	$\eta_1 = \dfrac{\lambda_1}{2\pi r_0} = b_1$	$\eta_2 = \dfrac{\lambda_2}{2\pi r_0} = \sqrt{2b_1 - 1}$
Length of one winding Eq. (1.6-3)	$l_1 = \dfrac{L_1}{2\pi r_0} = b_1$	$l_2 = \dfrac{L_2}{2\pi r_0} = b_1$
Steepness angle Eq. (1.6-4)	$\varphi_1 = 0$	$\cos s\varphi_2 = \dfrac{b_1 - 1}{b_1}$
The number of windings Eq. (1.6-5)	$w_1 = 1$	$w_2 = \dfrac{b_1}{\sqrt{2b_1 - 1}}$
Translational velocity Eq. (1.6-6)	$\beta_{1t} = \dfrac{V_{1t}}{c} = 0$	$\beta_{2t} = \dfrac{V_{2t}}{c} = \dfrac{\sqrt{2b_1 - 1}}{b_1}$
Rotational velocity Eq. (1.6-7)	$\beta_{1r} = \dfrac{V_{1r}}{c} = \dfrac{\sqrt{2b_1 - 1}}{b_1}$	$\beta_{2r} = \dfrac{V_{2r}}{c} = \dfrac{b_1 - 1}{b_1}$
Spiral velocity Eq. (1.6-8)	$\beta_1 = \dfrac{V_1}{c} = \dfrac{\sqrt{2b_1 - 1}}{b_1}$	$\beta_2 = \dfrac{V_2}{c} = 1$
Frequency Eq. (1.6-9)	$\delta_1 = \dfrac{f_1}{f_0} = \dfrac{\sqrt{2b_1 - 1}}{b_1^2}$	$\delta_2 = \dfrac{f_2}{f_0} = \dfrac{1}{b_1}$
Period Eq. (1.6-10)	$t_1 = T_1 f_0 = \dfrac{b_1^2}{\sqrt{2b_1 - 1}}$	$t_2 = T_2 f_0 = b_1$

The values of η_2, β_{2t} and β_{2r} correspond to the middle point a of toryx trailing string (Fig. 1.1.3). In Eq. (1.6-4b), $\cos s\varphi_2$ is the toryx trigonometric function. It relates to the trigonometric function $\cos\varphi_2$ of elementary mathematics as follows:

$$\cos s\varphi_2 = \cos\varphi_2 \quad (0 < \varphi_2 < 180^0) \tag{1.6-11}$$

$$\cos s\varphi_2 = 1/\cos\varphi_2 \quad (180^0 < \varphi_2 < 360^0) \tag{1.6-12}$$

The relative radius of toryx spherical boundary b is equal to:

$$b = \frac{r}{r_0} = 2b_1 - 1 \tag{1.6-13}$$

Table 1.6.2 provides a summary of derived equations for relative spacetime parameters of toryx leading and trailing strings as a function of the steepness angle of toryx trailing string φ_2.

Table 1.6.2. Equations for relative spacetime parameters of toryx leading and trailing strings as functions of the steepness angle of toryx trailing string φ_2.

Relative parameter	Leading string Eq. (a)	Trailing string Eq. (b)
Radius Eq. (1.6-14)	$b_1 = \dfrac{1}{1 - \cos s\varphi_2}$	$b_2 = \dfrac{\cos s\varphi_2}{1 - \cos s\varphi_2}$
Wavelength Eq. (1.6-15)	$\eta_1 = 0$	$\eta_2 = \sqrt{\dfrac{1 + \cos s\varphi_2}{1 - \cos s\varphi_2}}$
Length of one winding (Eq. 1.6-16)	$l_1 = \dfrac{1}{1 - \cos s\varphi_2}$	$l_2 = \dfrac{1}{1 - \cos s\varphi_2}$
The number of windings Eq. (1.6-17)	$w_1 = 1$	$w_2 = \dfrac{1}{\sin s\varphi_2}$
Translational velocity Eq. (1.6-18)	$\beta_{1t} = \dfrac{V_{1t}}{c} = 0$	$\beta_{2t} = \sin s\varphi_2$
Rotational velocity Eq. (1.6-19)	$\beta_{1r} = \sin s\varphi_2$	$\beta_{2r} = \cos s\varphi_2$
Spiral velocity Eq. (1.6-20)	$\beta_1 = \sin s\varphi_2$	$\beta_2 = 1$
Frequency Eq. (1.6-21)	$\delta_1 = \sin s\varphi_2 (1 - \cos s\varphi_2)$	$\delta_2 = 1 - \cos s\varphi_2$
Period Eq. (1.6-22)	$t_1 = \dfrac{1}{\sin s\varphi_2 (1 - \cos s\varphi_2)}$	$t_2 = \dfrac{1}{1 - \cos s\varphi_2}$

The relative radius of toryx spherical boundary b is equal to:

$$b = \frac{1 + \cos s\varphi_2}{1 - \cos s\varphi_2}$$

(1.6-23)

1.7 Toryx Law of Planetary Motion

The first strong indication that derived toryx basic spacetime equations were relevant to a real world comes from Eq. (1.6-8a). This equation establishes a relationship between the relative velocity of the toryx leading string β_1 and the relative radius of this string b_1 that is rewritten below:

$$\beta_1 = \frac{V_1}{c} = \frac{\sqrt{2b_1 - 1}}{b_1}$$

(1.7-1)

For the case when $b_1 \gg 1$, the above equation reduces to the form:

$$\beta_1 = \frac{V_1}{c} = \sqrt{\frac{2}{b_1}}$$

(1.7-2)

Let us show that Eq. (1.7-2) expresses the Kepler's third law of planetary motion.

The velocity of the toryx leading string V_1 is equal to:

$$V_1 = \frac{2\pi r_1}{T_1}$$

(1.7-3)

From Eqs. (1.6-1a), (1.7-2) and (1.7-3):

$$\frac{2\pi r_1}{T_1 c} = \sqrt{\frac{2r_0}{r_1}}$$

(1.7-4)

After squaring both parts of Eq. (1.7-4):

$$\frac{4\pi^2 r_1^2}{T_1^2 c^2} = \frac{2r_0}{r_1}$$

we obtain:

$$r_1^3 = \frac{r_0 c^2}{2\pi^2} T_1^2 \qquad (1.7\text{-}5)$$

Let k to be equal to:

$$k = \frac{r_0 c^2}{2\pi^2} \qquad (1.7\text{-}6)$$

From Eqs. (1.3-2) and (1.3-3), both r_0 and c are constant. Therefore, k is constant too. Consequently, Eqs. (1.7-5) and (1.7-6) yield Kepler's third law of planetary motion:

$$r_1^3 = k T_1^2 \qquad (1.7\text{-}7)$$

Both Eqs. (1.7-7) and (1.7-2) express Kepler's third law of planetary motion with only one difference: In Eq. (1.7-7) parameters are expressed in absolute units, while in Eq. (1.7-2) in relative units. Thus, Kepler's third law of planetary motion expressed by Eq. (1.7-2) is applied for large relative orbital radii ($b_1 \gg 1$) and can be treated as a particular case of the *toryx law of planetary motion* that is applied for the orbital radii extending from negative to positive infinity.

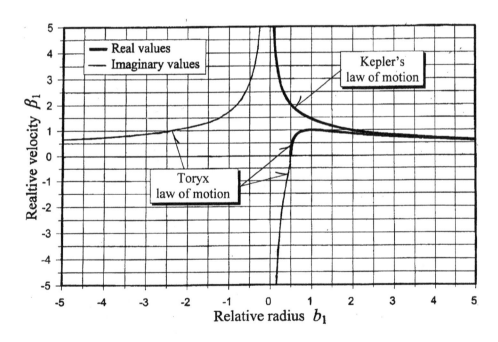

Figure 1.7. Toryx law of planetary motion versus Kepler's third law of planetary motion.

Figure 1.7 shows plots of Eqs. (1.7-1) and (1.7-2) expressing respectively the toryx law of planetary motion and Kepler's third law of planetary motion. The highlights of these plots are:

- The toryx law of motion is described by Eq. (1.7-1). It is applied to a range of b_1 extending from negative to positive infinity; within a range of b_1 extending from 0.5 to positive infinity the values of β_1 are expressed with real numbers, while within the remaining range of b_1 with imaginary numbers.

- Kepler's third law of planetary motion is described by Eq. (1.7-2). It is applied to the range of b_1 extending from zero to positive infinity with all values of β_1 expressed with real numbers.
- When $b_1 > 5$, the difference between the values of β_1 calculated based on the toryx law of planetary motion and Kepler's third law of planetary motion becomes small and it decrease as b_1 increases. As b_1 decreases from 5 to 2, this difference progressively increases.
- According to the Kepler's third law of planetary motion, as b_1 decreases from 2 to 0, β_1 sharply increases and approaches positive infinity $(+\infty)$. According to the toryx law of planetary motion, as b_1 decreases from 2 to 0. The value of β_1 initially slightly increases and then, after reaching its maximum value of 1 at $b_1 = 1$, it sharply decreases.

MAIN SUMMARY

- *Toryx is made up of a propagating double-circular leading string and a double-toroidal trailing string spiraling around and propagating synchronously with the leading string.*

- *Spacetime properties of a toryx are derived based on **three toryx basic genetic codes** (Exhibit 1.3 & 1.5) limiting its degree of freedom, so it becomes possible to establish a relationship between toryx parameters.*

- *The first two equations limit the degrees of freedom of its circular **leading string**, toroidal **trailing string** and its **eye radius**, while their lengths and radii can be expressed with real positive and negative numbers.*

- *The third equation requires the **spiral velocity of trailing string** to be constant and equal to the velocity of light expressed by the real positive number, while its translational and rotational velocity components can be subluminal, superluminal and can be expressed with positive, negative, real and imaginary numbers.*

- *The toryx basic genetic codes yield the **toryx law of planetary motion** for which the Kepler's third law of planetary motion is a particular case that is accurate when the orbital radius is much greater than the eye radius.*

2. FEATURES OF ABSTRACT MATHEMATICS OF A TORYX

CONTENTS

Equations describing the toryx spacetime parameters are mostly based on elementary math commonly taught in high schools. However, to satisfy the toryx spacetime postulates, it is necessary to modify several aspects of elementary math, including the definitions of zero, number line and elementary trigonometric functions. Also, unlike the elementary math that deals with stationary spiral elements, the toryx math considers the spiral elements in motion, explaining its name.

2.1 Infinility versus Elementary Zero

Conventionally, we use the elementary zero (0) in two ways. Firstly, we use it for counting of non-divisible entities. In an elementary number line (Fig. 2.1.1) it appears as an integer immediately preceding number one (1).

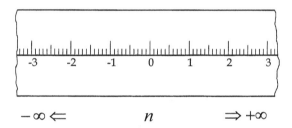

Figure 2.1.1. Elementary number line.

Secondly, we use zero to represent the absolute absence of any quantity and quality. Mathematically, the elementary zero (0) is equal to a ratio of one (1) to infinity (∞). The toryx math clearly separates two applications of zero described above. The zero is still considered as an integer for counting of non-divisible entities and still retains its old symbol (0). But, in application to the spacetime entities the zero is replaced with a quantity that is infinitely approaching to it. This quantity is called *infinility*, from the "infinite nil." (Notably, the term infinility is used in the toryx math instead of the known math term *infinitesimal*). In the toryx math, both infinity and infinility can be positive, negative, real and imaginary as shown below.

$$\text{Real infinity: } \pm\infty = \frac{1}{\pm 0}; \quad \text{Imaginary infinity: } \pm\infty i = \frac{1}{\pm 0i}$$

$$\text{Real infinility: } \pm 0 = \frac{1}{\pm\infty}; \quad \text{Imaginary infinility: } \pm 0i = \frac{1}{\pm\infty i}$$

Figure 2.1.2 shows symbolically positive and negative infinities $(\pm\infty)$ and also positive and negative infinility (± 0) as equal counterparts in respect to the positive and negative unities (± 1). In Nature, that range belongs to the **quantum vacuum** occupied by short-lived virtual toryces. Also shown in that figure is the **spacetime range** extending between the maximum value b_u and minimum value $b_u^{-1} = 1/b_u$. It is occupied by spacetime toryces.

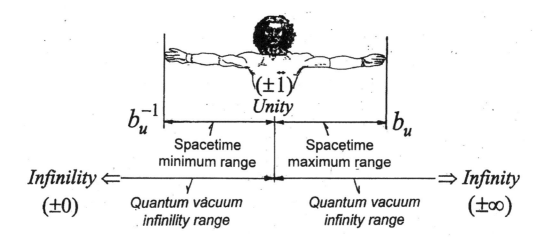

Figure 2.1.2. Infinity $(\pm\infty)$, infinility (± 0) and unity (± 1).

2.2 Spacetime Trigonometry

Definitions of elementary trigonometric functions are based on transformations of a right triangle as a function of the non-right angle φ_2 (Fig. 2.2.1):

$$\cos\varphi_2 = x \quad (0^0 < \varphi_2 < 360^0) \tag{2.2-1}$$

The main features of the transformations shown in Fig. 2.2.1 are:

- When the length of the hypotenuse of the triangles is equal to 1, the ranges of the lengths of its sides x and y are between 1 and -1.
- The triangles located in two left quadrants are the mirror images of the triangles located in two right quadrants.
- The triangles located in two bottom quadrants are the mirror images of the triangles located at two top quadrants.

In the spacetime trigonometry, the transformations of the right triangle are partially modified to satisfy the toryx spacetime postulates. Consequently, the spacetime trigonometric function $\cos s\varphi_2$ relates to the elementary trigonometric function $\cos\varphi_2$ as follows:

$$\cos s\varphi_2 = \cos\varphi_2 \quad (0 < \varphi_2 < 180^0) \tag{2.2-2}$$

$$\cos s\varphi_2 = 1/\cos\varphi_2 \quad (180^0 < \varphi_2 < 360^0) \tag{2.2-3}$$

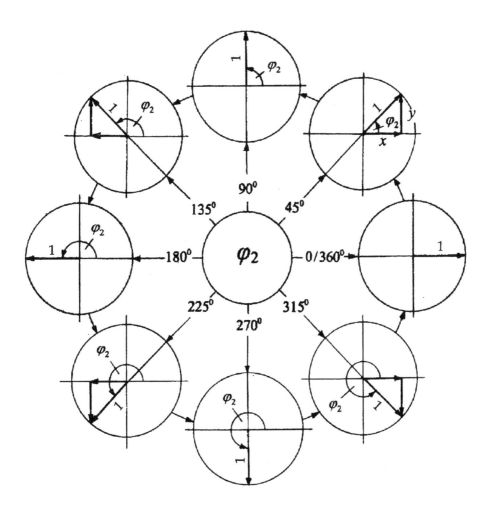

Figure 2.2.1. Transformations of a right triangle in elementary trigonometry.

The main features of the transformations shown in Fig. 2.2.2 are:

- When the angle φ_2 is between 0 and 180^0, the right triangles are the same as in elementary trigonometry. Thus, within this range of the angle φ_2 the elementary and spacetime trigonometry are based on the same principle.

- When the angle φ_2 is between 180 and 360^0, the right triangle becomes ***outverted***. Consequently, the length of its horizontal side x becomes greater than 1, while the length of the other side y is expressed with imaginary numbers.

- When the angle φ_2 approaches 270^0 from the angle smaller than 270^0, the length of its horizontal side x approaches real positive infinity $(+\infty)$, while the length of the other side y approaches imaginary positive infinity $(+\infty i)$.

- When the angle φ_2 approaches 270^0 from the angle greater than 270^0, the length of its horizontal side x approaches real negative infinity $(-\infty)$, while the length of the other side y approaches imaginary negative infinity $(-\infty i)$.

- When the angle φ_2 approaches 360^0 from the angle smaller than 360^0, the length of its horizontal side x approaches 1, while the length of the other imaginary side y approaches imaginary negative infinility $(-0i)$.

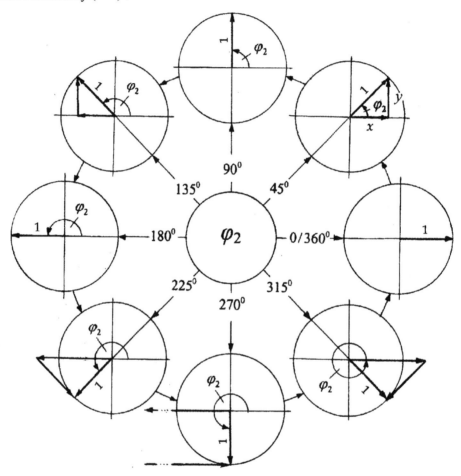

Figure 2.2.2. Transformations of a right triangle
in the spacetime trigonometry.

2.3 Toryx Number Lines

We consider below four kinds of toryx number lines that are directly related to the toryx parameters:

- *Toryx vorticity V number line*
- *Toryx reality R number line*
- *Toryx boundary B number line*
- *Toryx goldicity g number line*
- *Toryx string period ratio P number line.*

All five number lines are presented in the forms of circular diagrams in which the numbers V, R, B, g and P are expressed as functions of the steepness angle of toryx trailing string φ_2.

Toryx vorticity V number line - In the toryx vorticity V number line (Fig. 2.3.1), the real numbers V are equal to the ratio of toryx trailing string radius r_2 to the toryx leading string radius r_1 with an opposite sign. These numbers are extended clockwise along a circle from the real positive infinity $(+\infty)$ to the real negative infinity $(-\infty)$ as a function of the steepness angle of trailing string φ_2.

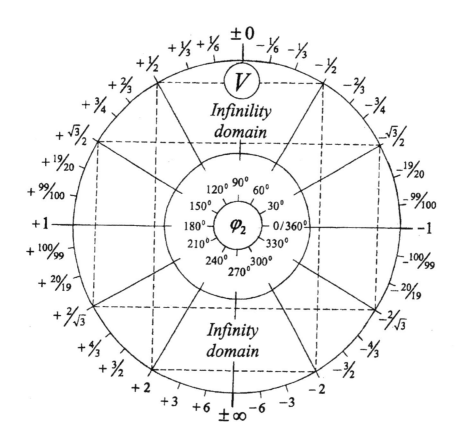

Figure 2.3.1. The toryx vorticity V number line.

$$V = -\frac{r_2}{r_1} = -\cos s\varphi_2 \qquad (2.3\text{-}1)$$

The toryx vorticity V number line is divided into two domains, the V *infinility domain* and the V *infinity domain*, occupying equal sectors of the circular number line.

- The V infinility domain occupies two top quadrants; it contains the values of V extending clockwise from the real positive unity $(+1)$ and passing through infinility (± 0) to the real negative unity (-1).

- The V infinity domain resides in two bottom quadrants; it contains the values of V extending counterclockwise from the real positive unity $(+1)$ and passing through infinity $(\pm\infty)$ to the real negative unity (-1).

In the toryx vorticity V number line, the real positive infinility $(+0)$ merges with real negative infinility (-0) at $\varphi_2 = 90^0$, while real negative infinity $(-\infty)$ merges with real positive infinity $(+\infty)$ at $\varphi_2 = 270^0$.

Symmetries between the toryx vorticity V numbers - There are two kinds of symmetries between the numbers V that belong to the four quadrants of circular diagram, the *inverse V-symmetry* and the *reverse V-symmetry*.

- In the inverse V-symmetry, the magnitudes of the numbers V located in the top quadrants are inversed (reciprocated) in respect to the magnitudes of the numbers V located in the bottom quadrants.

- In the reverse V-symmetry, the numbers V located in the right quadrants and the left quadrants have the same magnitudes but reversed signs.

Toryx reality R number line - In the toryx reality R number line (Fig. 2.3.2), the real and imaginary numbers R are equal to the ratio of toryx trailing string wavelength λ_2 to the circular length of toryx eye $l_0 = 2\pi r_0$. These numbers are extended counterclockwise along a circle from the real positive infinity $(+\infty)$ to the imaginary negative infinity $(-\infty i)$ as a function of the steepness angle of trailing string φ_2.

$$R = \frac{\lambda_2}{2\pi r_0} = \sqrt{\frac{1 + \cos s\varphi_2}{1 - \cos s\varphi_2}} \tag{2.3-2}$$

The toryx reality R number line is divided into two domains, the **R infinility domain** and the **R infinity domain,** occupying equal sectors of the circular number line.

- The R infinility domain occupies two left quadrants; it contains the values of R extending counterclockwise from the real positive unity $(+1)$ and passing through infinility $(+0/-0i)$ to the imaginary negative unity $(-i)$.

- The R infinity domain resides in two right quadrants; it contains the values of R extending clockwise from the real positive unity $(+1)$ and passing through real positive and imaginary negative infinities $(+\infty/-\infty i)$ to the imaginary negative unity $(-i)$.

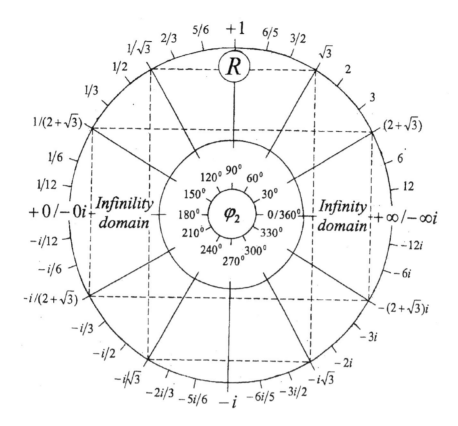

Figure 2.3.2. The toryx reality R number line.

In the toryx reality R number line, the real positive infinility ($+0$) merges with the imaginary negative infinility ($-0i$) at $\varphi_2 = 180^0$, while real positive infinity ($+\infty$) merges with imaginary negative infinity ($-\infty i$) at $\varphi_2 = 360^0$.

Symmetries between the toryx reality R numbers - There are two kinds of symmetries between the numbers R that belong to the four quadrants of circular diagram, the **inverse R-symmetry** and the **reverse reality R-symmetry**.

- In the inverse R-symmetry, the magnitudes of the numbers R located in the left quadrants are inversed (reciprocated) in respect to the magnitudes of the numbers R located in the right quadrants.
- In the reverse reality R-symmetry, the numbers R located in the top quadrants are real positive, while these numbers in the bottom quadrants are imaginary negative.

Toryx boundary B number line - In the toryx boundary B number line (Fig. 2.3.3), the real numbers B are equal to the ratio of the radius of toryx radius of spherical boundary r to the toryx eye radius r_0.

$$B = \frac{r}{r_0} = \frac{1 + \cos s\varphi_2}{1 - \cos s\varphi_2}$$

(2.3-3)

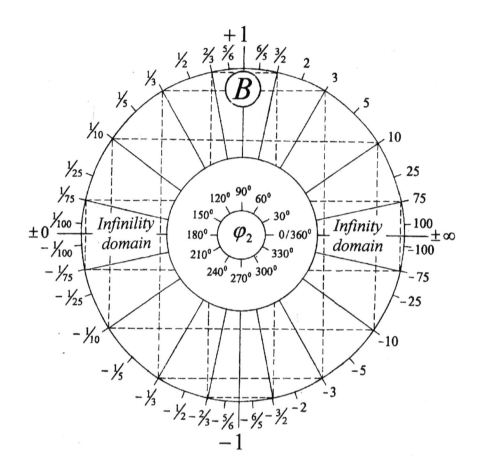

Figure 2.3.3. The toryx boundary B number line.

These numbers are extended counterclockwise along a circle from the real positive infinity $(+\infty)$ to the real negative infinity $(-\infty)$ as a function of the steepness angle of trailing string φ_2.

The toryx boundary B number line is divided into two domains, the ***B infinility domain*** and the ***B infinity domain***, occupying equal sectors of the circular number line.

- The B infinility domain occupies two left quadrants; it contains the values of B extending counterclockwise from the real positive unity $(+1)$ and passing through infinility $(+0/-0)$ to the real negative unity (-1).
- The B infinity domain resides in two right quadrants; it contains the values of B extending clockwise from the real positive unity $(+1)$ and passing through real positive and negative infinities $(+\infty/-\infty)$ to the real negative unity (-1).

In the toryx boundary B number line, real positive infinility $(+0)$ merges with real negative infinility (-0) at $\varphi_2 = 180^0$, while real positive infinity $(+\infty)$ merges with real negative infinity $(-\infty)$ at $\varphi_2 = 360^0$.

Symmetries between the toryx boundary B numbers - There are two kinds of symmetries between the numbers B that belong to the four quadrants of circular diagram, the *inverse B-symmetry* and the *reverse B-symmetry*.

- In the inverse B-symmetry, the magnitudes of the numbers B located in the left quadrants are inversed (reciprocated) in respect to the magnitudes of the numbers B located in the right quadrants.
- In the reverse B-symmetry, the numbers B located in the top and bottom quadrants have the same magnitudes but reversed signs.

Toryx goldicity g number line – There are two kinds of toryx goldicity number lines: *real* and *imaginary*. In the toryx real goldicity g number line (Fig. 2.3.4), the numbers g are related to the toryx radius of spherical boundary r, the eye radius r_0, the radius of leading string r_1 and the radius of trailing string r_2 of a toryx by the equation:

$$g = -\frac{r\,r_0}{r_1 r_2} = -\frac{1-\cos s^2\varphi_2}{\cos s\varphi_2} \tag{2.3-4}$$

When $g = \pm 1$, $\cos s\varphi_2$ relates to the **golden ratio** $\phi = (1+\sqrt{5})/2 = 1.618033989$ as shown in Table 2.3.

Table 2.3. $\cos s\varphi_2$ versus golden ratio ϕ.

Steepness angle of trailing string φ_2	$\cos s\varphi_2$	g
51.83^0	$+1/\phi$	-1
128.17^0	$-1/\phi$	$+1$
231.83^0	$-1/\phi$	-1
308.17^0	$+1/\phi$	$+1$

The toryx goldicity g number line is divided into four domains, two **g-infinity domains** and two **g-infinility domains**.

- The _g-infinity domains_ occupy top and bottom equal segments of the circular diagram. In both top and bottom g-infinity domains, the values of g extend clockwise from the real positive unity $(+1)$ and passing through infinity $(\pm\infty)$ to the real negative unity (-1).
- The _g-infinility domains_ occupy left and right equal segments of the circular diagram. In both left and right g -infinility domains, the values of g extend clockwise from the real negative unity (-1) and passing through infinility (± 0) to the real positive unity $(+1)$.

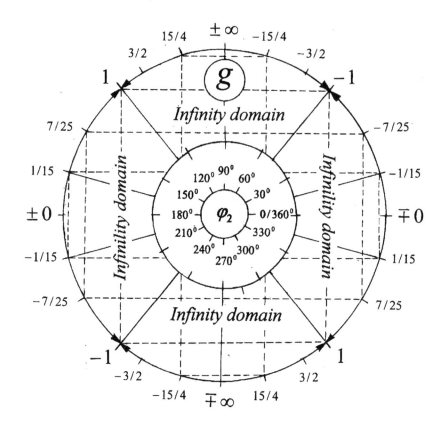

Figure 2.3.4. The toryx real goldicity g number line.

Symmetries between the toryx real goldicity g numbers - The numbers located in each of four quadrants of the g number line relate to the numbers of adjacent quadrants according to the *reverse g-symmetry*. It means that the numbers g located <u>in each of four quadrants</u> have the same magnitudes but reversed signs in respect to the numbers g located in their adjacent quadrants.

The toryx imaginary goldicity number $\breve{g}$ is related to the toryx real goldicity number g by the equation:

$$\breve{g} = -g\,i \tag{2.3-5}$$

Toryx string period ratio P number line – In the toryx string period ratio P number line (Fig. 2.3.5), the numbers P are related to the ratio of the periods of toryx trailing string T_2 and leading string T_1 by the equation:

$$P = \frac{T_2}{T_1} = \sin s\varphi_2 \tag{2.3-5}$$

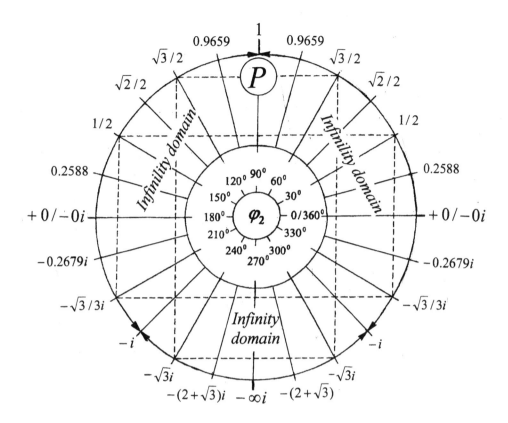

Figure 2.3.5. The toryx string period ratio P number line.

The toryx string period ratio P number line is divided into three domains, two ***P-infinility domains*** and one ***P-infinity domains***.

- Two *P*-infinility domains occupy top left and right equal segments of the circular diagram. In both left and right top segments, the values of P extend respectively counter-clockwise and clockwise from the real positive unity $(+1)$ and passing through infinility $(+0/-0i)$ to the imaginary negative unity $(-i)$. The numbers P located in the left and right top segments are equal to one another.
- The *P*-infinity domain occupy bottom left and right segments of the circular diagram. In both left and right bottom segments, the values of P extend from the imaginary negative unity $(-i)$ and pass through imaginary negative infinity $(-\infty i)$ to the imaginary negative unity $(-i)$. The numbers P located in the left and right bottom segments are equal to one another.

2.4. Toryx Parameters in Number Lines & Trigonometry

The numbers V of the toryx vorticity V number line (Fig. 2.3.1) relate to the toryx parameters by the equation:

$$V = -\frac{r_2}{r_1} = -\frac{b_2}{b_1} = -\frac{b_1 - 1}{b_1} = -\frac{b-1}{b+1} = -\beta_{2r} = -\cos s\varphi_2 \qquad (2.4\text{-}1)$$

The numbers R of the toryx reality R number line (Fig. 2.3.2) relate to the toryx parameters by the equation:

$$R = \eta_2 = \sqrt{b} = \sqrt{2b_1 - 1} = b_1\beta_{2t} = \sqrt{\frac{1 + \cos s\varphi_2}{1 - \cos s\varphi_2}} \qquad (2.4\text{-}2)$$

The numbers B of the toryx boundary B number line (Fig. 2.3.3) relate to the toryx parameters by the equation:

$$B = b = \frac{r}{r_0} = 2b_1 - 1 = \frac{1 + \cos s\varphi_2}{1 - \cos s\varphi_2} \qquad (2.4\text{-}3)$$

The numbers g of the toryx real goldicity g number line (Fig. 2.3.4) are related to the toryx parameters by the equation:

$$g = -\frac{r\,r_0}{r_1\,r_2} = -\frac{b}{b_1\,b_2} = -\frac{4b}{b^2 - 1} = -\frac{2b_1 - 1}{b_1(b_1 - 1)} = -\frac{\beta_{2t}^2}{\beta_{2r}} = -\frac{1 - \cos s^2\varphi_2}{\cos s\varphi_2} \qquad (2.4\text{-}4)$$

When $g = \pm 1$, the toryx parameters relate to the golden ratio ϕ as shown in Table 2.4.

Table 2.4. Relationships between toryx parameters when $g = \pm 1$.

φ_2	$\cos s\varphi_2$	g	V	b_2	b_1	b
51.83^0	$+1/\phi$	-1	$-1/\phi$	ϕ	$1 + \phi$	$1 + 2\phi$
128.17^0	$-1/\phi$	$+1$	$+1/\phi$	$-1/\phi^2$	$1/\phi$	$2/\phi - 1$
231.83^0	$-1/\phi$	-1	$+1/\phi$	$-1/\phi$	$1/\phi^2$	$2/\phi^2 - 1$
308.17^0	$+1/\phi$	$+1$	$-1/\phi$	$-(1 + \phi)$	$-\phi$	$-(1 + 2\phi)$

The numbers P of the toryx string period ratio P number line relate to the toryx parameters by the equation:

$$P = \frac{T_2}{T_1} = = \frac{t_2}{t_1} = \frac{\sqrt{2b_1 - 1}}{b_1} = \frac{1}{w_2} = \beta_1 = \beta_{1r} = \beta_{2t} = \sin s\varphi_2 \qquad (2.4\text{-}5)$$

Figure 2.4 shows the application of the spiral spacetime math for the calculation of the relative velocities of the toryx trailing string corresponding to the middle point of the trailing string as its steepness angle φ_2 increases from 0 to 360^0. In each right triangle of velocities of trailing string one side represents the relative translational velocity β_{2t} and the other side the relative rotational velocity β_{2r}, while its hypotenuse represents the relative spiral velocity $\beta_2 = 1$.

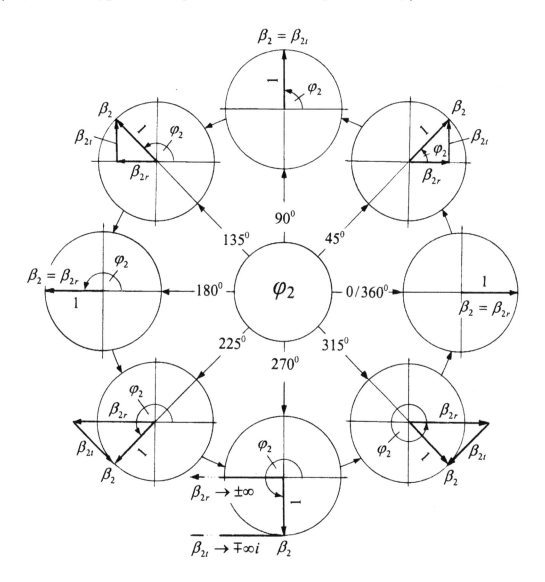

Figure 2.4. Transformations of right triangle representing vectors of the relative velocities of toryx trailing string β_2, β_{2t} and β_{2r} corresponding to the middle point a of trailing string (Fig. 1.1.3).

MAIN SUMMARY

Based on three basic toryx spacetime postulates, it is necessary to modify four commonly-accepted aspects of elementary mathematics.

- *Conventional zero is replaced with **infinility** (± 0) that is an inverse of **infinity** ($\pm \infty$).*

- *Trigonometric **cosine** function between 180^0 and 360^0 is replaced with its inverse value.*

- *Several **circular number lines** are used to provide a possibility to express a full range of toryx parameters and to reveal symmetrical relationships between them.*

3. CLASSIFICATION OF TORYCES

3.1 Main Groups and Subgroups of Toryces

Depending on the toryx vorticity V and reality R numbers the toryces are divided into four *main groups* and eight *subgroups* as shown in Tables 3.1.1 and 3.1.2.

Main groups of toryces - The toryces of the main groups are described below and their features are summarized in Table 3.1.1.

Real negative toryces $(0^0 < \varphi_2 < 90^0)$ – The real negative toryces are located in the top right quadrants of the circular diagrams. They are called "real negative" because their reality R numbers are real and their vorticities numbers V are negative.

Real positive toryces $(90^0 < \varphi_2 < 180^0)$ – The real positive toryces are located in the top left quadrants of the circular diagrams. They are called "real positive" because their reality R numbers are real and their vorticities numbers V are positive.

Imaginary positive toryces $(180^0 < \varphi_2 < 270^0)$ – The imaginary toryces are located in the bottom left quadrants of the circular diagrams. They are called "imaginary positive" because their reality R numbers are imaginary and their vorticities numbers V are positive.

Imaginary negative toryces $(270^0 < \varphi_2 < 360^0)$ – The imaginary negative toryces are located in the bottom right quadrants of the circular diagrams. They are called "imaginary negative" because their reality R numbers are imaginary and their vorticities numbers V are negative.

Table 3.1.1. The realities R, the vorticities V and the mid relative rotational velocities β_{2r} of trailing strings of toryces of main groups.

Toryces of main groups	φ_2	R	V	β_{2r}
Real negative	0^0 - 90^0	Real	$(-)$	< 1 (subluminal)
Real positive	90^0 - 180^0	Real	$(+)$	< 1 (subluminal)
Imaginary positive	180^0 - 270^0	Imaginary	$(+)$	> 1 (superluminal)
Imaginary negative	270^0 - 360^0	Imaginary	$(-)$	>1 (superluminal)

Subgroups of toryces - Within each main group, the toryces are further divided into two *subgroups* as shown in Table 3.1.2.

Table 3.1.2. Toryces of main groups and subgroups.

Toryces of main groups	Toryces of subgroups	
	Name	φ_2
Real negative	E^-	$0^0 < \varphi_2 < 60^0$
	A^-	$60^0 < \varphi_2 < 90^0$
Real positive	A^+	$90^0 < \varphi_2 < 120^0$
	E^+	$120^0 < \varphi_2 < 180^0$
Imaginary positive	$\breve{E}^+$	$180^0 < \varphi_2 < 240^0$
	$\breve{A}^+$	$240^0 < \varphi_2 < 270^0$
Imaginary negative	$\breve{A}^-$	$270^0 < \varphi_2 < 300^0$
	$\breve{E}^-$	$300^0 < \varphi_2 < 360^0$

3.2 Basic Toryces

The toryces located at the borderlines of the subgroups of toryces ($\varphi_2 = 60^0, 120^0, 240^0$ & 300^0) are called the ***basic toryces***. Figures 3.2.1 – 3.2-4 show the relationships between parameters of basic toryces.

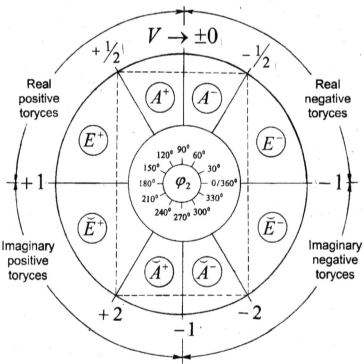

Figure 3.2.1. Relationships between the vorticities V of basic toryces.

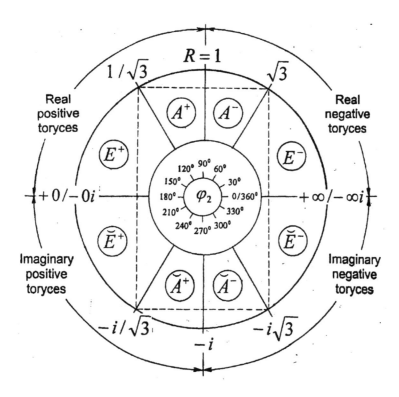

Figure 3.2.2. Relationships between the realities R of basic toryces.

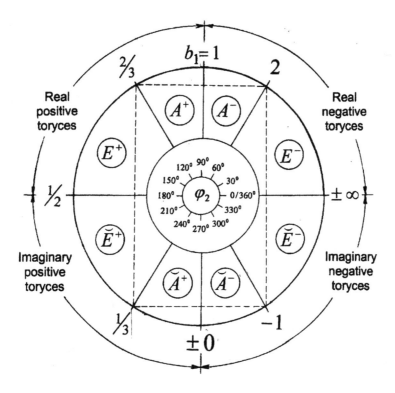

Figure 3.2.3. Relationships between the relative radii of leading strings b_1
of basic toryces.

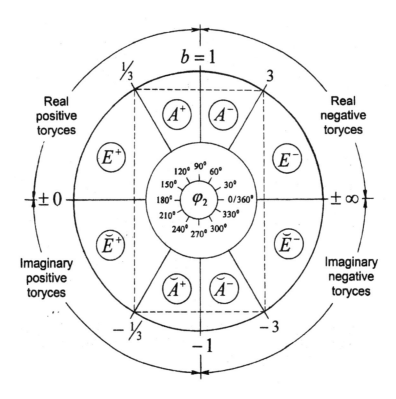

Figure 3.2.4. Relationships between the relative radii of spherical boundaries b of basic toryces.

3.3 Relationships between Parameters of Adjacent Toryces

Adjacent toryces of main groups - Table 3.3.1 shows relationships between the vorticities V and the realities R of adjacent toryces of main groups.

Table 3.3.1. Relationships between the vorticity V and the reality R of adjacent toryces of main groups.

Adjacent toryces		Eqs. (a)	Eqs. (b)
Reality-polarized negative toryces	$\breve{E}^-$ & E^- Eq. (3.3-1)	$\breve{V}_E^- = 1/V_E^-$	$\breve{R}_E^- = \pm i R_E^-$
Reality-polarized positive toryces	$\breve{E}^+$ & E^+ Eq. (3.3-2)	$\breve{V}_E^+ = 1/V_E^+$	$\breve{R}_E^+ = \pm i R_E^+$
Vorticity-polarized real toryces	A^+ & A^- Eq. (3.3-3)	$V_A^+ = -V_A^-$	$R_A^+ = 1/R_A^-$
Vorticity-polarized imaginary toryces	$\breve{A}^+$ & $\breve{A}^-$ Eq. (3.3-4)	$\breve{V}_A^+ = -\breve{V}_A^-$	$\breve{R}_A^+ = 1/\breve{R}_A^-$

Table 3.3.2 shows relationships between the relative radii of leading strings b_1 and spherical boundaries b of adjacent toryces of main groups.

Table 3.3.2. Relationships between the relative radii of leading string b_1 and spherical boundary b of adjacent toryces.

Adjacent toryces		Eqs. (a)	Eqs. (b)
Reality-polarized negative toryces	$\breve{E}^- \& E^-$ Eq. (3.3-5)	$\breve{b}_{1E}^- = 1 - b_{1E}^-$	$\breve{b}_E^- = -b_E^-$
Reality-polarized positive toryces	$\breve{E}^+ \& E^+$ Eq. (3.3-6)	$\breve{b}_{1E}^+ = 1 - b_{1E}^+$	$\breve{b}_E^+ = -b_E^+$
Vorticity-polarized real toryces	$A^+ \& A^-$ Eq. (3.3-7)	$b_{1A}^+ = \dfrac{b_{1A}^-}{2b_{1A}^- - 1}$	$b_A^+ = \dfrac{1}{b_A^-}$
Vorticity-polarized imaginary toryces	$\breve{A}^+ \& \breve{A}^-$ Eq. (3.3-8)	$\breve{b}_{1A}^+ = \dfrac{\breve{b}_{1A}^-}{2b_{1A}^- - 1}$	$\breve{b}_A^+ = \dfrac{1}{\breve{b}_A^-}$

Adjacent toryces of subgroups - Table 3.3.3 shows the relationships between the vorticities V of adjacent toryces of subgroups.

Table 3.3.3. Relationships between vorticities V of adjacent toryces of subgroups.

Adjacent toryces of subgroups		Equations
Real negative toryces	$A^- \& E^-$	$V_A^- + V_E^- = -1 \quad (3.3\text{-}9)$
Real positive toryces	$A^+ \& E^+$	$V_A^+ + V_E^+ = +1 \quad (3.3\text{-}10)$
Imaginary positive toryces	$\breve{A}^+ \& \breve{E}^+$	$\dfrac{1}{V_A^+} + \dfrac{1}{V_E^+} = +1 \quad (3.3\text{-}11)$
Imaginary negative toryces	$\breve{A}^- \& \breve{E}^-$	$\dfrac{1}{V_A^-} + \dfrac{1}{V_E^-} = -1 \quad (3.3\text{-}12)$

Table 3.3.4 summarizes relationships between the relative radii of leading strings b_1 and spherical boundaries b of adjacent toryces of subgroups.

Table 3.3.4. Relationships between the relative radii of leading string b_1
and spherical boundary b of adjacent toryces of subgroups.

Toryces of subgroups		Eqs. (a)	Eqs. (b)
Negative toryces	A^- & E^- Eq. (3.3-13)	$b_{1A}^- = \dfrac{b_{1E}^-}{b_{1E}^- - 1}$	$b_A^- = \dfrac{b_E^- + 3}{b_E^- - 1}$
Positive toryces	A^+ & E^+ Eq. (3.3-14)	$b_{1A}^+ = \dfrac{b_{1E}^+}{3b_{1E}^+ - 1}$	$b_A^+ = \dfrac{1 - b_E^+}{1 + 3b_E^+}$

3.4 Mutually-Polarized and Self-Polarized Toryces

During each cycle of a trailing string its instant translational velocity β_{2t} and rotational velocity β_{2r} vary. Still, according to Eq. (1.5-3), their geometrical sum remains constant and equals to the relative velocity of light ($\beta_2 = 1$). As shown in Fig. 3.4.1, the instant translational velocity at each point of trailing string is proportional to the distance of that point from the toryx center.

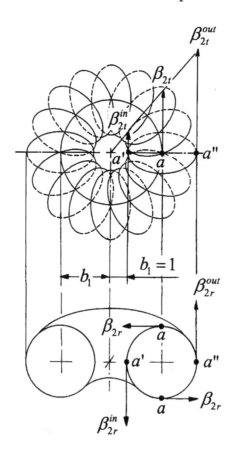

Figure 3.4.1. Relative instant velocities of trailing string.

Table 3.4 shows equations for relative instant inner and outer peripheral translational and rotational velocities of trailing strings.

Table 3.4. Relative instant inner and outer peripheral velocities of trailing string.

Inner velocities At the point a' (Fig. 3.4.1)		**Outer velocities** at the point a'' (Fig. 3.4.1)	
$$\beta_{2t}^{in} = \frac{\sqrt{2b_1 - 1}}{b_1^2}$$	(3.4-1)	$$\beta_{2t}^{out} = \frac{(2b_1 - 1)^{3/2}}{b_1^2}$$	(3.4-2)
$$\beta_{2r}^{in} = \frac{\sqrt{b_1^4 - 2b_1 + 1}}{b_1^2}$$	(3.4-3)	$$\beta_{2r}^{out} = \frac{\sqrt{b_1^4 - (2b_1 - 1)^3}}{b_1^2}$$	(3.4-4)

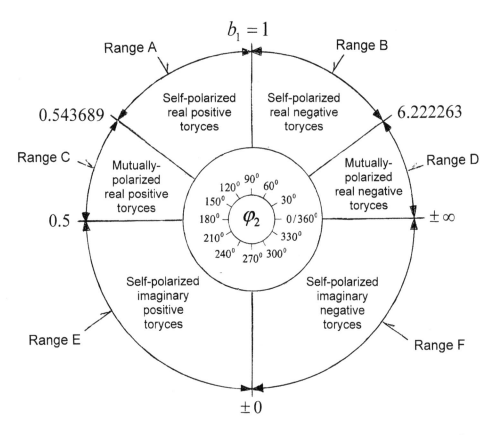

Figure 3.4.2. The ranges of the relative radius of leading string b_1 of mutually-polarized and self-polarized toryces.

Figure 3.4.2 shows the ranges of the relative radii of leading string b_1 of mutually-polarized and self-polarized toryces. In the mutually polarized toryces (ranges C and D), the instant translational and rotational velocities of trailing stings are subluminal during all of their cycles. Consequently, both these velocities are expressed with real numbers. In the self-polarized toryces (ranges A, B, E and F), the instant translational velocities of trailing string are superluminal during at least some parts of their cycles. Consequently, their instant rotational velocities are expressed with imaginary numbers during those parts of cycles.

Figure 3.4.3 shows the variation of the relative instant translational and rotational veloci-ties, β_{2t}^i and β_{2r}^i for the case when $b_1 = 2$. Notably, between 0.25 and 0.75 portions of the cycle of trailing string, β_{2t}^i exceeds velocity of light, while β_{2r}^i is expressed with imaginary numbers.

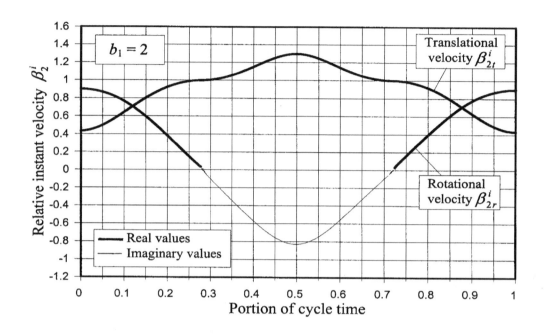

Figure 3.4.3. Variation of instant velocities β_{2t}^i and β_{2r}^i of trailing string within one cycle of trailing string.

MAIN SUMMARY

Based on their reality R and vorticity V, the toryces are divided into four main groups:

- *Real negative toryces*
- *Real positive toryces*
- *Imaginary positive toryces*
- *Imaginary negative toryces.*

Each main group of toryces is divided into two subgroups.

There are symmetrical relationships between parameters of toryces that belong to different groups and subgroups.

The toryces are further divided into two groups:

- *Mutually-polarized toryces*
- *Self-polarized toryces.*

4. Trends of Toryx Parameters

Contents

The toryces having the spacetime parameters described in the previous Chapters are called the *basic toryces*. Presented below are the plots of equations expressing the basic toryx parameters as functions of the steepness angle of trailing string φ_2.

4.1 Radii of Leading & Trailing Strings

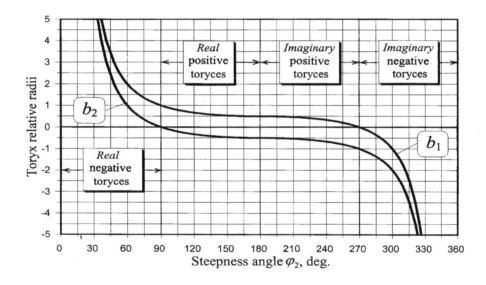

φ_2	$360^0/0^0$	90^0	180^0	270^0
b_1	$-\infty/+\infty$	$+1$	$+\frac{1}{2}$	$+0/-0$
b_2	$-\infty/+\infty$	$+0/-0$	$-\frac{1}{2}$	-1

Figure 4.1. Trends of the relative radii of leading and trailing strings, b_1 and b_2.

4.2 Radius of Spherical Boundary

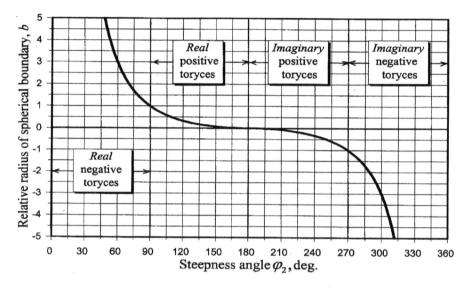

φ_2	$360^0/0^0$	90^0	180^0	270^0
b	$-\infty/+\infty$	$+1$	$+0/-0$	-1

Figure 4.2. Trend of the toryx relative radius of spherical boundary b.

4.3 Wavelength of Trailing String

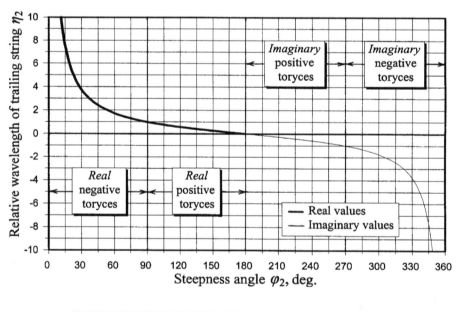

φ_2	$360^0/0^0$	90^0	180^0	270^0
η_2	$-\infty i/+\infty$	$+1$	$+0/-0i$	$-i$

Figure 4.3. Trend of the relative wavelength of trailing string η_2.

4.4 The Number of Windings of Trailing String

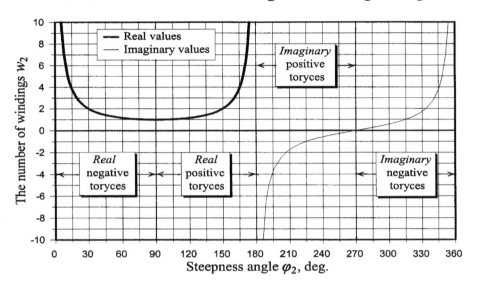

φ_2	$360^0 / 0^0$	90^0	180^0	270^0
w_2	$+\infty i / +\infty$	$+1$	$+\infty / -\infty i$	$-0i / +0i$

Figure 4.4. Trend of the number of windings of trailing string w_2.

4.5 Translational Velocity of Trailing String

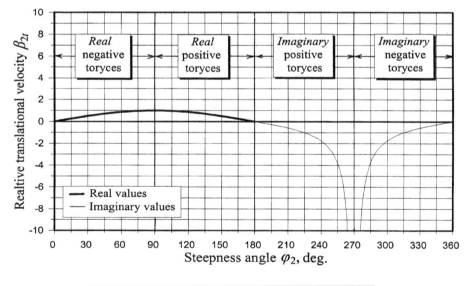

φ_2	$360^0 / 0^0$	90^0	180^0	270^0
β_{2t}	$-0i / +0$	$+1$	$+0 / -0i$	$-\infty i$

Figure 4.5. Trend of the relative translational velocity of trailing string β_{2t}.

4.6 Rotational Velocity of Trailing String

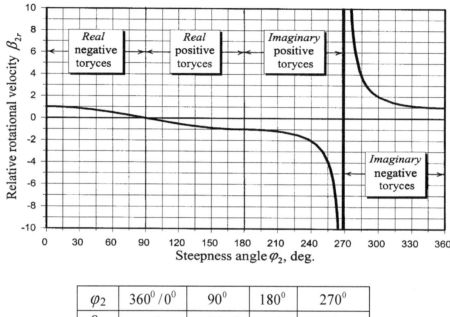

φ_2	$360^0/0^0$	90^0	180^0	270^0
β_{2r}	$+1$	$+0/-0$	-1	$-\infty/+\infty$

Figure 4.6. Trend of the relative rotational velocity of trailing string β_{2r}.

4.7 Frequency of Leading String

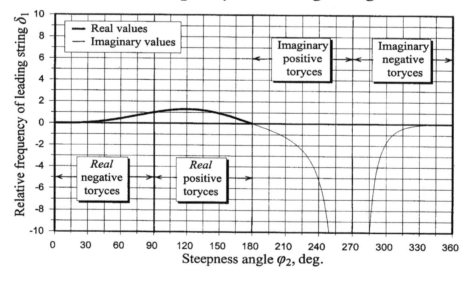

φ_2	$360^0/0^0$	90^0	180^0	270^0
δ_1	$-0i/+0$	$+1$	$+0/-0i$	$-\infty i$

Figure 4.7. Trend of the relative frequency of leading string δ_1.

4.8 Frequency of Trailing String

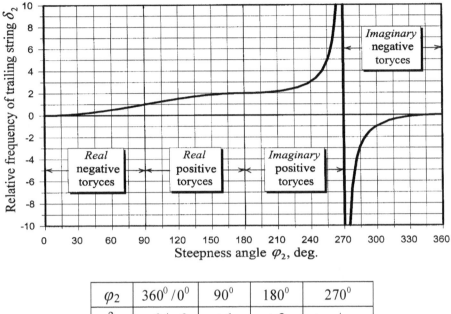

φ_2	$360^0/0^0$	90^0	180^0	270^0
δ_2	$-0/+0$	$+1$	$+2$	$+\infty/-\infty$

Figure 4.8. Trend of the relative frequency of trailing string δ_2.

4.9 Period of Leading String

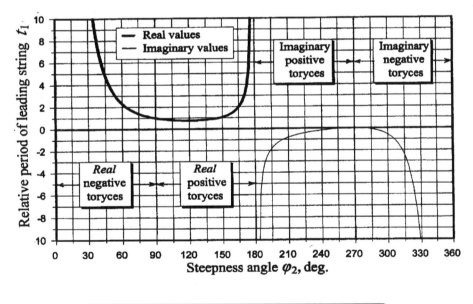

φ_2	$360^0/0^0$	90^0	180^0	270^0
t_1	$-\infty i/+\infty$	$+1$	$+\infty/-\infty i$	$-0i$

Figure 4.9. Trend of the relative period of leading string t_1.

4.10 Period of Trailing String

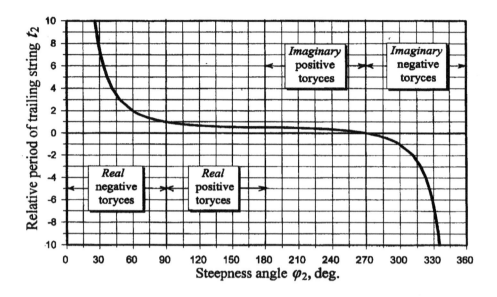

φ_2	$360^0/0^0$	90^0	180^0	270^0
t_2	$-\infty/+\infty$	$+1$	$+\frac{1}{2}$	$+0/-0$

Figure 4.10. Trend of the relative period of trailing string t_2.

MAIN SUMMARY

Trends of spacetime properties of toryces are expressed as a function of steepness angle of toryx trailing string extending from 0^0 to 360^0 degrees, corresponding to the radius of toryx leading string extending from negative to positive infinity.

5. INVERSION OF TORYCES

CONTENTS

5.1 Inversion of Toryx Leading & Trailing Strings

As the radius of toryx leading string b_1 decreases from positive to negative infinity, the leading string and trailing string undergo through spacetime transformations described below.

Inversion of leading string – Visualize the toryx leading string in the form of an extremely thin and narrow circular ribbon with the relative radius b_1 (Fig. 5.1.1). When $b_1 > 0$; the outer color of the circular ribbon is assumed to be black while its inner color is white. The leading string remains **outverted** until b_1 reduces to positive infinility $(+0)$. When b_1 approaches infinility (± 0), the leading string becomes **inverted**. So, when $b_1 < 0$, the outer color of leading string appears white, while its inner color looks black.

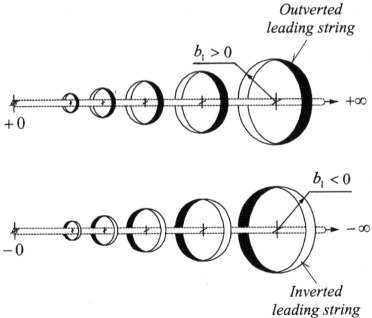

Fig. 5.1.1. Inversion of a leading string.

Inversion of a trailing string - Visualize the toryx trailing string in the form of an extremely thin and narrow toroidal ribbon with the relative radius b_2 (Fig. 5.1.2). When the relative radius of leading string $b_1 > 1$ the radius of trailing string $b_2 > 1$. For that case, the outer color of toroidal ribbon is assumed to be black, while its inner color white. The trailing string remains *outverted* until b_2 reduces to positive infinity $(+0)$. When $b_1 = 1$, b_2 approaches infinility (± 0) and the trailing string becomes *inverted*. So, when $b_2 < 1$, the outer color of leading string appears white, while its inner color looks black.

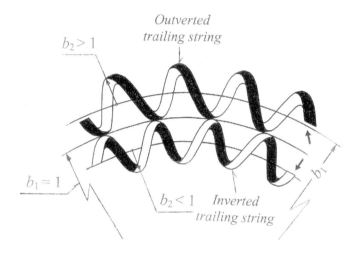

Fig. 5.1.2. Inversion of a trailing string.

Inversion of a spherical boundary - Visualize an extremely thin spherical boundary with the relative radius b (Fig. 5.1.3).

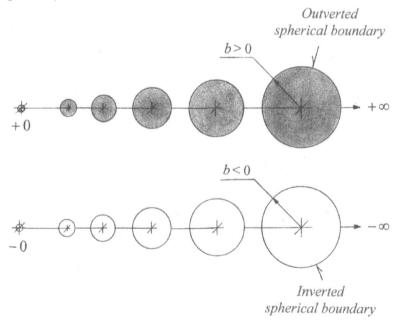

Fig. 5.1.3. Inversion of a spherical boundary.

When $b > 0$, the outer color of the spherical boundary is assumed to be grey and its inner color is assumed to be white. The spherical boundary remains **outverted** until b reduces to positive infinility $(+0)$. When b approaches infinility (± 0), the spherical boundary becomes **inverted**. So, when $b < 0$, the outer color of spherical boundary appears white, while its inner color becomes grey.

5.2 Metamorphoses of Toryces

Figure 5.2 shows that as the radius of the toryx leading string decreases from positive to negative infinity, the steepness angle of trailing string φ_2 increases from 0^0 to 360^0, while the **toryx vorticity** V extends equally from **Unity** (± 1) towards both **Infinility** (± 0) and **Infinity** $(\pm \infty)$.

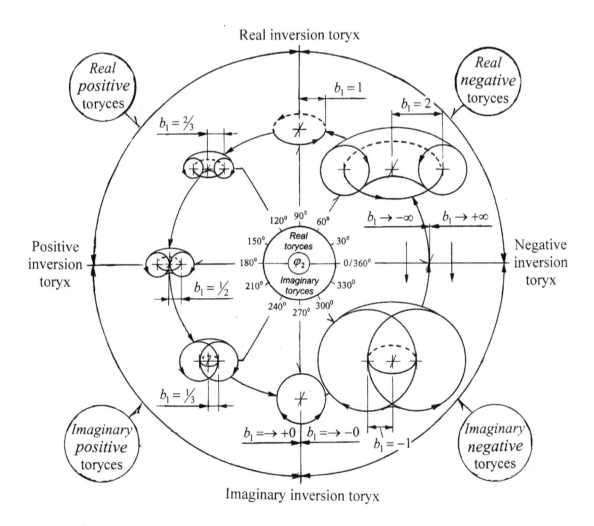

Figure 5.2. Metamorphoses of toryx leading and trailing strings as a function of the steepness angle of trailing string φ_2.

Four kinds of *inversion toryces* are located at the boundaries of four quadrants of the circular diagram.

- **Negative inversion toryx** $(\varphi_2 \to 0^0/360^0)$ - At this point, the toryx leading string, trailing string and its wavelength become inverted, and the toryx appears as two parallel lines separated by the distance equal to the diameter of real inversion string.

- **Real inversion toryx** $(\varphi_2 \to 90^0)$ – At this point, the toryx trailing string becomes inverted $(b_1 \to +1, b_2 \to \pm 0, \eta_2 = 1)$, and the toryx appears as a circle with the relative radius $b_1 \to +1$.

- **Positive inversion toryx** $(\varphi_2 \to 180^0)$ – At this point, the wavelength of toryx trailing string becomes inverted $(b_1 \to +0.5, b_2 \to -0.5, \eta_2 \to +0/-0i)$, and the toryx appears as an extreme case of a *spindle torus* with the inner parts of its windings touching one another.

- **Imaginary inversion toryx** $(\varphi_2 \to 270^0)$ – At this point, the toryx leading string becomes inverted $(b_1 \to \pm 0, b_2 \to -1, \eta_2 = -i)$, and the toryx appears as a circle with the relative radius approaching -1. The circle is located at the plane perpendicular to the plane of the real inversion string.

Table 5.2 shows extreme relative parameters of inversion toryces.

Table 5.2. Extreme relative parameters of inversion toryces.

Inversion toryces	φ_2	b	b_1	b_2	η_2	w_2	β_{2t}	β_{2r}	δ_1	δ_2
Negative inversion toryx	$0^0/360^0$	$-\infty$ $+\infty$	$-\infty$ $+\infty$	$-\infty$ $+\infty$	$-\infty i$ $+\infty$	$+\infty i$ $+\infty$	$-0i$ $+0$	$+0$ -0	$-0i$ $+0$	-0 $+0$
Real inversion toryx	90^0	-	-	$+0$ -0	-	-	-	$+0$ -0	-	-
Positive inversion toryx	180^0	$+0$ -0	$+0$ -0	-	$+0$ $-0i$	$+\infty$ $-\infty i$	$+0$ $-0i$	$-\infty$ $+\infty$	$+0$ $-0i$	-
Imaginary inversion toryx	270^0	-	-	-	-	$-0i$ $+0i$	$-\infty i$	-	$-\infty i$	$+\infty$ $-\infty$

5.3 Transformations of Real Negative Toryces

Real negative toryces (Fig. 5.3) belong to the top right quadrant of the circular diagram shown in Fig. 5.2. Trailing strings of these toryces are wound counter-clockwise outside of the real inversion toryx. As φ_2 increases, both b_1 and b_2 decrease, so that the trailing string appears like a conventional torus.

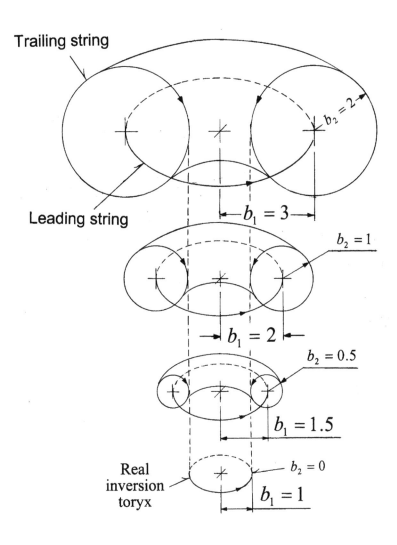

Range	φ_2	b	b_1	b_2	η_2	w_2	β_{2t}	β_{2r}
From	0^0	$+\infty$	$+\infty$	$+\infty$	$+\infty$	$+\infty$	$+0$	1.0
To	90^0	1.0	1.0	$+0$	1.0	1.0	1.0	$+0$

Figure 5.3. Transformations of real negative toryces.

5.4 Transformations of Real Positive Toryces

Real positive toryces (Fig. 5.4) belong to the top left quadrant of the circular diagram shown in Figure 5.2. Within this range the trailing string is inverted, so that its windings are now wound clockwise inside the real inversion toryx. As φ_2 increases, b_1 decreases, while the negative value of b_2 increases. Consequently, the toryx appears as an inverted toroidal spiral.

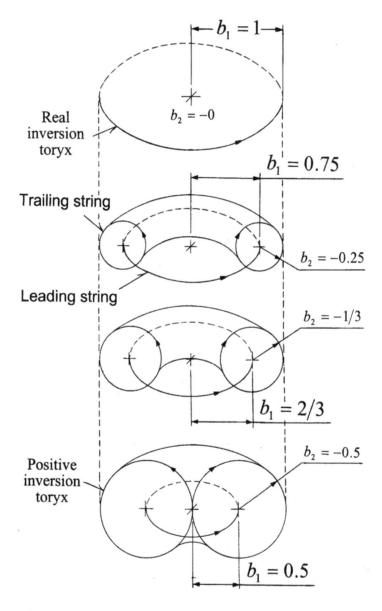

Range	φ_2	b	b_1	b_2	η_2	w_2	β_{2t}	β_{2r}
From	90^0	1.0	1.0	- 0	1.0	1.0	1.0	- 0
To	180^0	+0	0.5	-0.5	+0	$+\infty$	+0	-1.0

Figure 5.4. Transformations of real positive toryces.

5.5 Transformations of Imaginary Positive Toryces

Real positive toryces (Fig. 5.5) belong to the bottom left quadrant of the circular diagram shown in Figure 5.2. As φ_2 increases, b_1 decreases while negative values of b_2 increase. Within this range, the opposite parts of windings of trailing string intersect with one another like in a spindle torus.

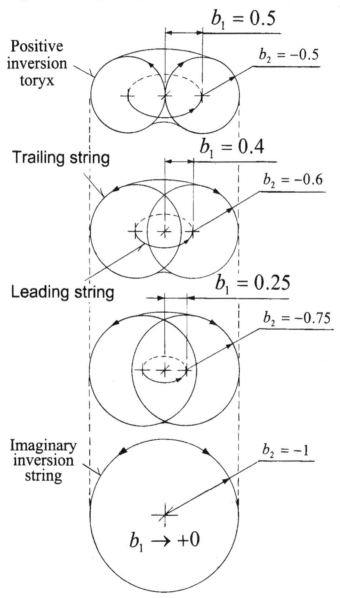

Range	φ_2	b	b_1	b_2	η_2	w_2	β_{2t}	β_{2r}
From	180^0	-0	0.5	-0.5	$-0i$	$-\infty i$	$-0i$	-1.0
To	270^0	-1.0	$+0$	-1.0	$-i$	$-0i$	$-\infty i$	$-\infty$

Figure 5.5. Transformations of imaginary positive toryces.

5.6 Transformations of Imaginary Negative Toryces

Imaginary negative toryces (Fig. 5.6) belong to the bottom right quadrant of the circular diagram shown in Figure 5.2. Here the leading string becomes inverted. As φ_2 increases, the negative values of b_1 increase, the negative values of b_2 also increase. Within this range the toryx windings are located outside of the imaginary inversion toryx.

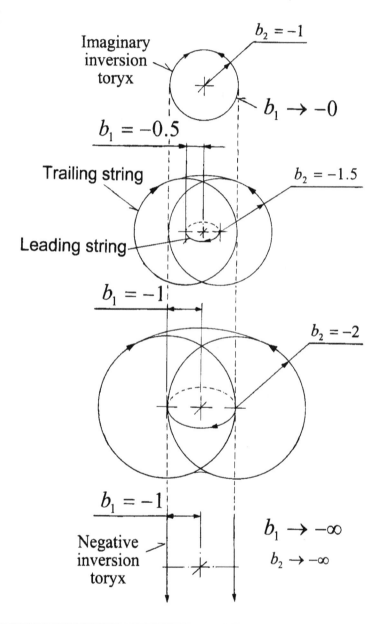

Range	φ_2	b	b_1	b_2	η_2	w_2	β_{2t}	β_{2r}
From	270^0	-1.0	-0	-1.0	$-i$	$+0i$	$-\infty i$	$+\infty$
To	360^0	$-\infty$	$-\infty$	$-\infty$	$-\infty i$	$+\infty i$	$-0i$	1.0

Figure 5.6. Transformations of imaginary negative toryces.

5.7 Summary of Toryx Transformations

Figure 5.7 summarizes transformation of toryces by showing together the metamorphoses of toryx spherical boundary, leading string and trailing string with the respective relative radii b, b_1 and b_2 as the steepness angle of trailing string φ_2 increases from 0^0 to 360^0.

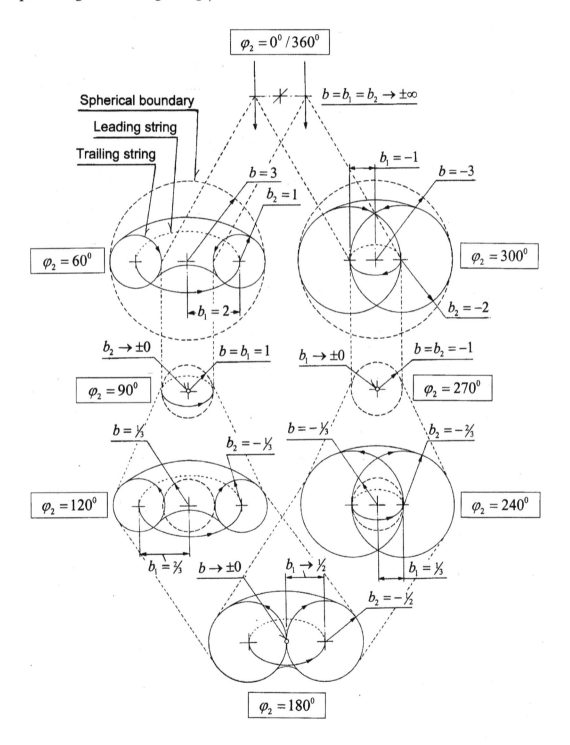

Figure 5.7. Transformations of toryx spherical boundary, leading string and trailing string as a function of the steepness angle of trailing string φ_2.

- When $\varphi_2 \to +0^0$, all three radii b, b_1 and b_2 approach positive infinity $(+\infty)$.

- Within the range of $\varphi_2 (+0^0 < \varphi_2 < 90^0)$, b, b_1 and b_2 decrease as φ_2 increases.

- When $\varphi_2 \to 90^0$, $b = b_1 = 1$ and b_2 approaches positive infinility $(+0)$; the ***trailing string becomes inverted*** and b_2 becomes negative.

- Within the range of $\varphi_2 (90^0 < \varphi_2 < 180^0)$, b_1 and b continue to decrease, while negative values of b_2 increase starting from negative infinility (-0).

- When $\varphi_2 \to 180^0$, $b_1 = \frac{1}{2}$, $b_2 = -\frac{1}{2}$ and b approaches positive infinility $(+0)$; the ***toryx spherical boundary becomes inverted*** and b becomes negative.

- Within the range of $\varphi_2 (180^0 < \varphi_2 < 270^0)$, b_1 continues to decrease, negative values of b_2 continue to increase, while negative values of b increase starting from negative infinility (-0).

- When $\varphi_2 \to 270^0$, $b = b_2 = -1$, while b_1 approaches positive infinility $(+0)$; the ***leading string becomes inverted*** and b_1 becomes negative.

- Within the range of $\varphi_2 (270^0 < \varphi_2 < 360^0)$, negative values of b_1 increase starting from negative infinility (-0), while negative values of b and b_2 continue to increase.

- When $\varphi_2 \to 360^0$, all three radii b, b_1 and b_2 approach negative infinity $(-\infty)$.

Table 5.7. Inversion points of toryx leading string,
trailing strings and spherical boundary.

Toryx component	Relative radius	Steepness angle
Spherical boundary	$b \to \pm\infty$	$\varphi_2 \to 0^0/360^0$
	$b \to \pm 0$	$\varphi_2 \to 180^0$
Leading string	$b_1 \to \pm\infty$	$\varphi_2 \to 0^0/360^0$
	$b_1 \to \pm 0$	$\varphi_2 \to 270^0$
Trailing string	$b_2 \to \pm\infty$	$\varphi_2 \to 0^0/360^0$
	$b_2 \to \pm 0$	$\varphi_2 \to 90^0$

Table 5.7 summarizes the inversion steepness angles of trailing string and corresponding radii of toryx leading string, trailing string and spherical boundary.

5.8 Golden Toryces

Figure 5.8 and Table 2.4 show cross-sections and parameters of four golden toryces corresponding to the toryx goldicity $g = \pm 1$.

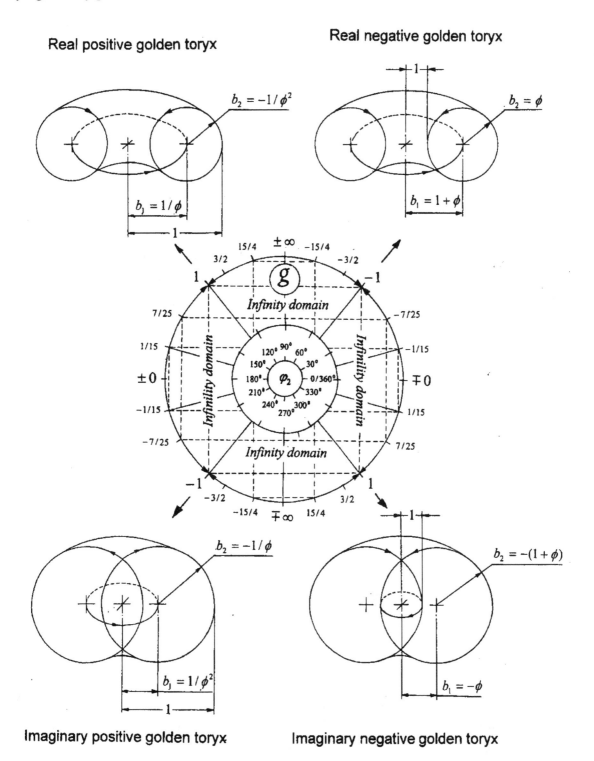

Figure 5.8 Golden toryces corresponding to the toryx goldicity $g = \pm 1$.

5.9 Minimum and Maximum Values of Toryx Parameters

As shown in Fig. 5.9, the minimum and maximum values of the toryx parameters are at the steepness angles of the toryx trailing string $\varphi_2 \to 0/360^0, 90^0, 180^0 \ \& \ 270^0$.

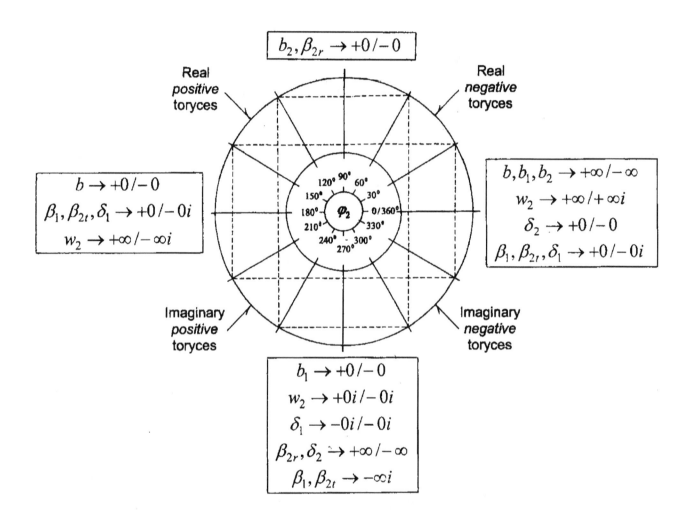

Figure 5.9. Minimum and maximum values of the toryx parameters.

5.10 Toryces Versus Antitoryces

Antitoryces have the same structures as their respective toryces. Also the same are their geometrical parameters. The differences between them is in the inversion states of their strings and spherical membranes that are inverted in respect to one another. Table 5.10 shows the appearances of leading and trailing strings of toryces and antitoryces.

Table 5.10 String appearances of toryces and antitoryces.

Toryx & antitoryx		Figures
Leading string	Toryx	
	Antitoryx	
Trailing string	Toryx	
	Antitoryx	

MAIN SUMMARY

As the radius of toryx leading string decreases from positive to negative infinity and the steepness angle of toryx trailing string φ_2 extending from 0^0 to 360^0 degrees, the toryx undergoes through four topological inversions:

- *Toryx trailing string inverts at* $\varphi_2 = 90^0$
- *Wavelength of toryx trailing string and spherical boundary invert at* $\varphi_2 = 180^0$
- *Toryx leading string inverts at* $\varphi_2 = 270^0$
- *All toryx components invert at* $\varphi_2 = 0^0 / 360^0$.

Antitoryces have the same structures as their respective toryces. Also the same are their geometrical parameters. The difference between them is in the inversion states of their strings and spherical boundaries that are inverted in respect to one another.

PART 2

Applied
Mathematics
of a Toryx

6. QUANTUM STATES & FINE STRUCTURE OF TORYCES

CONTENTS

6.1 Definition of Excitation & Oscillation of Toryces

Toryces change their dimensions in quantum steps by *excitation* and *oscillation process*. As shown in Figure 6.1, during excitation of a toryx the radius of toryx leading string r_1 increases, while its eye radius r_0 remains constant. During oscillation of a toryx, its radius of leading string r_1 and its eye radius r_0 change with the same rate.

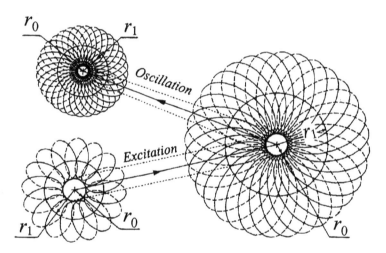

Figure 6.1. Oscillation of a toryx versus excitation of a toryx.

6.2 Quantization Equations for Excited Toryces

The derived quantization equations for excited toryces are based on three proposed limitations of degrees of freedom of real negative *lambda, harmonic* and *golden toryces* shown in Exhibit 6.2.

Exhibit 6.2. Limitations of degrees of freedom of excited toryces.

The relative radius of leading string of real negative toryx b_1 is equal to:		
Lambda toryx	**Harmonic toryx**	**Golden toryx**
$b_1 = z = 2(n\Lambda)^m$ (6.2-1)	$b_1 = 2 + n$ (6.2-2)	$b_1 = 2 + n/\phi$ (6.2-3)

In Exhibit 6.2:

$m = 0, 1, 2, \ldots$ toryx exponential excitation quantum states

$n = 0, 1, 2, \ldots$ toryx linear excitation quantum states

$\Lambda = 137$, ***toryx quantization constant*** that can be expressed as a function of three integers 1, 2 and 3:

$$(1 \times 2)^2 + (2 \times 2)^2 + (2 \times 3)^2 + (3 \times 3)^2 = 137.$$

Table 6.2.1 shows quantization equations for relative radii of leading strings and spherical boundries of excited ***lambda toryces***.

Table 6.2.1. Quantization equations for relative radii of toryx leading strings and spherical boundaries of excited lambda toryces.

Lambda toryx	Relative radius of leading string		Relative radius of spherical boundary	
$E^-_{m,n,q}$	$b^-_{1E} = z$	(6.2-4)	$b^-_E = 2z - 1$	(6.2-5)
$\breve{E}^-_{m,n,q}$	$\breve{b}^-_{1E} = 1 - z$	(6.2-6)	$\breve{b}^-_E = 1 - 2z$	(6.2-7)
$E^+_{m,n,q}$	$b^+_{1E} = \dfrac{z}{2z - 1}$	(6.2-8)	$b^+_E = \dfrac{1}{2z - 1}$	(6.2-9)
$\breve{E}^+_{m,n,q}$	$\breve{b}^+_{1E} = \dfrac{1 - z}{1 - 2z}$	(6.2-10)	$\breve{b}^+_E = \dfrac{1}{1 - 2z}$	(6.2-11)
$A^-_{m,n,q}$	$b^-_{1A} = \dfrac{z}{z - 1}$	(6.2-12)	$b^-_A = \dfrac{z + 1}{z - 1}$	(6.2-13)
$A^+_{m,n,q}$	$b^+_{1A} = \dfrac{z}{z + 1}$	(6.2-14)	$b^+_A = \dfrac{z - 1}{z + 1}$	(6.2-15)
$\breve{A}^-_{m,n,q}$	$\breve{b}^-_{1A} = \dfrac{1}{1 - z}$	(6.2-16)	$\breve{b}^-_A = \dfrac{1 + z}{1 - z}$	(6.2-17)
$\breve{A}^+_{m,n,q}$	$\breve{b}^+_{1A} = \dfrac{1}{1 + z}$	(6.2-18)	$\breve{b}^+_A = \dfrac{1 - z}{1 + z}$	(6.2-19)

The first subscripts in the symbols of toryces are as shown below:

$$E^-_{m,n,q}, \quad \breve{A}^+_{m,n,q}, \quad E^-_{H,n,q}, \quad A^+_{G,n,q}$$

m is the toryx exponential excitation quantum state in lambda toryces
H is standing for harmonic toryces
G is standing for golden toryces.

The second subscript indicates the toryx linear excitation quantum state n.
The third subscript indicates the oscillation quantum states q (see Section 6.3).
The superscript indicates the sign of toryx vorticity V and, in some cases, their values.
The "smile" sign at the top of a symbol indicates an imaginary toryx.

Table 6.2.2 show quantization equations for relative radii of leading strings and spherical boundaries of excited harmonic toryces.

Table 6.2.2. Quantization equations for relative radii of toryx leading strings and spherical boundaries of excited harmonic toryces.

Harmonic toryx	Relative radius of leading string		Relative radius of spherical boundary	
$E_{H,n,q}^{-}$	$b_{1E}^{-} = 2+n$	(6.2-20)	$b_{E}^{-} = 3+2n$	(6.2-21)
$\breve{E}_{H,n,q}^{-}$	$\breve{b}_{1E}^{-} = -(1+n)$	(6.2-22)	$\breve{b}_{E}^{-} = -(3+2n)$	(6.2-23)
$E_{H,n,q}^{+}$	$b_{1E}^{+} = \dfrac{2+n}{3+2n}$	(6.2-24)	$b_{E}^{+} = \dfrac{1}{3+2n}$	(6.2-25)
$\breve{E}_{H,n,q}^{+}$	$\breve{b}_{1E}^{+} = \dfrac{1+n}{3+2n}$	(6.2-26)	$\breve{b}_{E}^{+} = -\dfrac{1}{3+2n}$	(6.2-27)
$A_{H,n.q}^{-}$	$b_{1A}^{-} = \dfrac{2+n}{1+n}$	(6.2-28)	$b_{A}^{-} = \dfrac{3+n}{1+n}$	(6.2-29)
$A_{H,n.q}^{+}$	$b_{1A}^{+} = \dfrac{2+n}{3+n}$	(6.2-30)	$b_{A}^{+} = \dfrac{1+n}{3+n}$	(6.2-31)
$\breve{A}_{H,n.q}^{-}$	$\breve{b}_{1A}^{-} = -\dfrac{1}{1+n}$	(6.2-32)	$\breve{b}_{A}^{-} = -\dfrac{3+n}{1+n}$	(6.2-33)
$\breve{A}_{H,n,q}^{+}$	$\breve{b}_{1A}^{+} = \dfrac{1}{3+n}$	(6.2-34)	$\breve{b}_{A}^{+} = -\dfrac{1+n}{3+n}$	(6.2-35)

Table 6.2.3 shows quantization equations for relative radii of leading strings and spherical boundaries of excited golden toryces.

Table 6.2.3. Quantization equations for relative radii of toryx leading strings and spherical boundaries of excited golden toryces.

Golden toryx	Relative radius of leading string	Relative radius of spherical boundary
$E_{G,n,q}^{-}$	$b_{1E}^{-} = 2 + n/\varphi$ (6.2-36)	$b_{E}^{-} = 3 + 2n/\varphi$ (6.2-37)
$\breve{E}_{G,n,q}^{-}$	$\breve{b}_{1E}^{-} = -(1 + n/\varphi)$ (6.2-38)	$\breve{b}_{E}^{-} = -(3 + 2n/\varphi)$ (6.2-39)
$E_{G,n,q}^{+}$	$b_{1E}^{+} = \dfrac{2+n}{3+2n/\varphi}$ (6.2-40)	$b_{E}^{+} = \dfrac{1}{3+2n/\varphi}$ (6.2-41)
$\breve{E}_{G,n,q}^{+}$	$\breve{b}_{1E}^{+} = \dfrac{1+n/\varphi}{3+2n/\varphi}$ (6.2-42)	$\breve{b}_{E}^{+} = -\dfrac{1}{3+2n/\varphi}$ (6.2-43)
$A_{G,n.q}^{-}$	$b_{1A}^{-} = \dfrac{2+n/\varphi}{1+n/\varphi}$ (6.2-44)	$b_{A}^{-} = \dfrac{3+n/\varphi}{1+n/\varphi}$ (6.2-45)
$A_{G,n.q}^{+}$	$b_{1A}^{+} = \dfrac{2+n/\varphi}{3+n/\varphi}$ (6.2-46)	$b_{A}^{+} = \dfrac{1+n/\varphi}{3+n/\varphi}$ (6.2-47)
$\breve{A}_{G,n.q}^{-}$	$\breve{b}_{1A}^{-} = -\dfrac{1}{1+n/\varphi}$ (6.2-48)	$\breve{b}_{A}^{-} = -\dfrac{3+n/\varphi}{1+n/\varphi}$ (6.2-49)
$\breve{A}_{G,n,q}^{+}$	$\breve{b}_{1A}^{+} = \dfrac{1}{3+n/\varphi}$ (6.2-50)	$\breve{b}_{A}^{+} = -\dfrac{1+n/\varphi}{3+n/\varphi}$ (6.2-51)

Basic toryces - For particular cases of basic toryces when $m = 0$ and $n = 0$, the equations shown in Tables 6.2-1 - 6.2-3 yield the quantum states of harmonic toryces located at the borderlines between polarized toryces, so the adjacent harmonic toryces have the same structures and properties.

Spectral gap systems – The spectral gap systems are divided into two groups, *gapped system* and *gapless system*. Equation (6.2-1) and equations shown in Table (6.2.1) describe quantum states of relative radii of toryx leading strings and spherical boundaries of excited lambda toryces of the gapped system. Equations (6.2-2), (6.2-3) and equations shown in Tables (6.2-2) and (6.2-3) describe quantum states of relative radii of toryx leading strings and spherical boundaries of excited harmonic and golden toryces of the gapless system.

6.3 Quantization Equations for Oscillated Toryces

The *toryx oscillation factor* Q_q describes the degree of oscillation of a toryx. This factor is expressed as a function of the *toryx oscillation quantum state* q as shown in Exhibit 6.3.

Exhibit 6.3. Limitations of degrees of freedom of oscillated toryces.

The toryx oscillation factor Q_q is equal to:

$$q = 0: \quad Q_0 = 1$$

$$q = 1, 2, 3..: \quad Q_q = 3\left(\frac{\Lambda}{2(q-1)}\right)^{q-1} \qquad (6.3\text{-}1)$$

Figure 6.3 shows a plot of the toryx oscillation factor Q_q as a function of the toryx oscillation quantum states q calculated from Eq. (6.3-1).

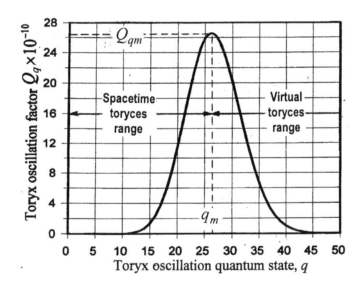

Figure 6.3.1. Toryx oscillation factor Q_q as a function of the toryx oscillation quantum state q.

At $q = q_m$, the toryx oscillation factor Q_q reaches the maximum value Q_{qm} as shown below.

$$q_m = 1 + \frac{\Lambda}{2e} = 26.199742$$

$$Q_{qm} = 3e^{\Lambda/2e} = 2.637728 \times 10^{11} \qquad (6.3\text{-}2)$$

where e is the natural logarithm base.

When $q > q_m$, the toryx oscillation factor Q_q sharply decreases and at $q = 70.589986$ its magnitude reduces to 1. After that as q continues to increase and approaches infinity, Q_q decreases and approaches infinility.

6.4 Fine Structure of Toryces

Fine structure of a toryx is formed by a ***standing wave*** propagating over the toryx ***inner spherical boundary*** (Fig. 6.4) with the radius r_3 described by the equation:

$$r_3 = \sqrt{r_1^2 - r_2^2} \qquad (6.4\text{-}1)$$

The relative radius of toryx inner spherical boundary b_3 is equal to:

$$b_3 = \eta_2 = \sqrt{2b_1 - 1} \qquad (6.4\text{-}2)$$

The perimeter $2\pi r_3$ of toryx inner spherical boundary is equal to the wavelength λ_2 of toryx trailing string:

$$2\pi r_3 = \lambda_2 \qquad (6.4\text{-}3)$$

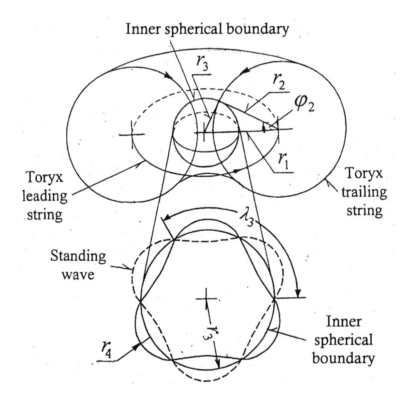

Figure 6.4. Fine structure of a real negative toryx.

The spiral length of trailing string of toryx standing wave L_4 is related to its radius r_4 and wavelength λ_3 by the equation:

$$L_4 = \sqrt{\lambda_3^2 + (2\pi r_4)^2} \qquad (6.4\text{-}4)$$

To define the radius r_4 and frequency f_3 of toryx standing wave, it is necessary to introduce the limitations of its degrees of freedom shown in Exhibit 6.4.

Exhibit 6.4. Limitation of degrees of freedom of toryx standing wave.

The aspect ratio of toryx standing wave $\lambda_3/2r_4$ is equal to the toryx quantization constant Λ:

$$\frac{\lambda_3}{2r_4} = \Lambda = \text{const.} \qquad (6.4\text{-}5)$$

The frequency of toryx standing wave f_3 is equal to the frequency of toryx trailing string f_2:

$$f_3 = f_2 \qquad (6.4\text{-}6)$$

From Eqs. (6.4-4) and (6.4-5), the spiral length of toryx standing wave L_4 is equal to:

$$L_4 = \lambda_3 \frac{\alpha_s^{-1}}{\Lambda} \qquad (6.4\text{-}7)$$

where α_S^{-1} is the *spacetime inverse fine-structure constant* related to the spacetime quantization constant Λ by the equation:

$$\alpha_S^{-1} = \sqrt{\Lambda^2 + \pi^2} \qquad (6.4\text{-}8)$$

Notably, Eq. (6.4-8) is similar to the equation proposed by T.J. Burger as an approximation of the inverse fine structure constant. For the case when $\Lambda = 137$, this equation yields the value of $\alpha_S^{-1} = 136.036015720$ that is very close to the experimental value of the inverse fine structure constant $\alpha^{-1} = 136.035999074$ provided by 2011 CODATA.

6.5 Radius of Toryx String Thickness

Exhibit 6.5 and Fig. 6.5 show the assumed value of the relative radius of toryx string thickness b_s.

Exhibit 6.5. Relative radius of toryx string thickness.

The relative radius of toryx string thickness b_s is equal to:

$$b_s = b_u^{-1} = 1/(2Q_{qm} - 1) = 1.895571 \times 10^{-12} \qquad (6.5\text{-}1)$$

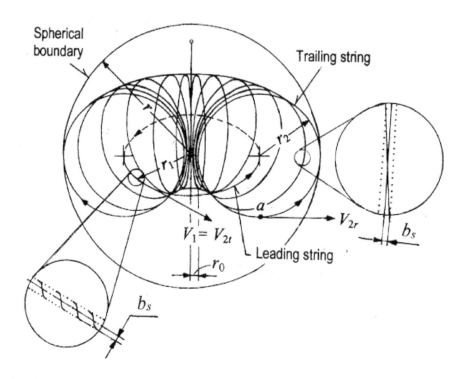

Figure 6.5. The relative radii of leading and trailing toryx string thicknesses b_s .

MAIN SUMMARY

- *Toryces may exist at the excitation quantum states allowing only the radius of toryx leading string to change, while its eye radius remains constant - see Exhibit 6.2.*

- *Toryces may exist at the oscillation quantum states allowing both the radius of the toryx leading string and its eye radius to change in the same proportion - see Exhibit 6.3.*

- *The spacetime parameters of the standing wave are defined by the limitations of its degrees of freedom shown in Exhibit 6.4.*

- *The relative radius of the toryx string thickness b_s is equal to the ultimate minimum relative radius of toryx spherical membrane b_u^{-1} as described by Eq.6.5-1 and shown in Fig. 6.5.*

7. Toryx "Rosetta Stone"

This theory is based on an assumption that both space and time are universal terms that can be discovered not only by human beings but also by all intelligent inhabitants of the Multiverse. However, physical terms, such as mass, charge, force, energy, magnetic induction, etc. are purely human-made inventions based on the theories developed by their scientists. To make that theory universally understandable, it expresses the properties of entities of the Multiverse in both spacetime and physical terms, similarly to the "Rosetta stone."

7.1 Relationships between Physical & Spacetime Properties of Toryces

The following physical constants are used for describing toryx physical parameters:

c = velocity of light in vacuum r_e = classical electron radius
e_0 = elementary charge μ_B = Bohr magneton
h = Planck constant μ_N = nuclear magneton
m_e = electron mass α_s = spacetime fine structure constant
m_p = proton mass ε_0 = electric constant.

Derivation of relationships between toryx physical and spacetime parameters are based on two postulates shown in Exhibit 7.2.

Exhibit 7.1. Basic relationships between toryx physical and spacetime parameters.

- The relative toryx charge e_t / e_0 is equal to the toryx vorticity V:

$$\frac{e_t}{e_0} = V \qquad (7.1\text{-}1)$$

- The relative toryx gravitational mass m_g / m_e is proportional to the absolute value of the toryx vorticity $|V|$:

$$\frac{m_g}{m_e} = Q_q |V| \qquad (7.1\text{-}2)$$

Table 7.1.1 shows derived relationships between toryx physical and spacetime parameters.

Table 7.1.1. Relationships between toryx relative physical and spacetime parameters.

Toryx relative parameters	Equations	
Charge	$$\frac{e_t}{e_0} = -\frac{r_2}{r_1} = -\frac{b_1-1}{b_1} = V$$	(7.1-3)
Gravitational mass	$$\frac{m_g}{m_e} = Q_q\left|\frac{r_2}{r_1}\right| = Q_q\left|\frac{b_1-1}{b_1}\right| = Q_q|V|$$	(7.1-4)
Inertial mass	$$\frac{m_{ti}}{m_e} = Q_q\frac{2r_2}{r} = Q_q\frac{2(b_1-1)}{2b_1-1} = Q_q\frac{R^2-1}{R^2}$$	(7.1-5)
Real magnetic moment	$$\frac{\mu_t}{\mu_B} = \pm\alpha_s\frac{(b_1-1)\sqrt{2b_1-1}}{2Q_q b_1} = \pm\alpha_s\frac{VR}{2Q_q}$$	(7.1-6a)
Imaginary magnetic moment	$$\frac{\breve{\mu}_t}{\mu_B} = \pm\alpha_s i\frac{(b_1-1)\sqrt{2b_1-1}}{2Q_q b_1} = \pm\alpha_s i\frac{VR}{2Q_q}$$	(7.1-6b)
Matter energy	$$\frac{E_m}{m_e c^2}V = Q_q\frac{b_1-1}{b_1} = -Q_q V$$	(7.1-7)
Field energy	$$\frac{E_f}{m_e c^2} = \frac{Q_q}{b_1} = Q_q\delta_2$$	(7.1-8)
Total matter & field energy	$$\frac{E_t}{m_e c^2} = 1$$	(7.1-9)
Total kinetic & potential energy	$$\frac{E_t}{m_e c^2} = -Q_q\frac{(b_1-1)}{b_1^2} = Q_q V\delta_2$$	(7.1-9)
Real spacetime intensity	$$\frac{E_s}{m_e c^2} = I = Q_q g = -Q_q\frac{2b_1-1}{(b_1-1)b_1} = Q_q\frac{(R\delta_2)^2}{V}$$	(7.1-10a)
Imaginary spacetime intensity	$$\frac{\breve{E}_s}{m_e c^2} = \breve{I} = Q_q\breve{g} = iQ_q\frac{2b_1-1}{(b_1-1)b_1} = iQ_q\frac{(R\delta_2)^2}{V}$$	(7.1-10b)
Density	$$\rho_t\frac{2\pi^2 r_0^3}{m_e} = \frac{Q_q}{b_1(b_1-1)^2}\left|\frac{b_1-1}{b_1}\right| = \frac{Q_q\delta_2^3}{|V|}$$	(7.1-11)
Modulus of elasticity	$$B_t\frac{2\pi^2 r_0^3}{m_e c^2} = Q_q\left|\frac{b_1-1}{b_1}\right|\frac{2b_1-1}{b_1^3(b_1-1)^2} = \frac{Q_q R^2\delta_2^5}{|V|}$$	(7.1-12)

7.2. Examples of Graphical Presentations of Toryx Parameters

The graphs shown below are made for the case when the toryx oscillation factor $Q_q = 1$.

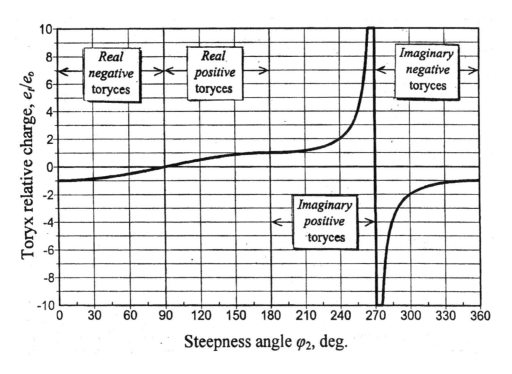

Fig. 7.2.1. Toryx relative charge.

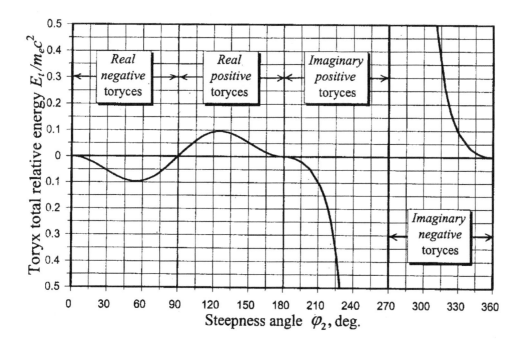

Fig. 7.2.2. Toryx total relative kinetic & potential energy.

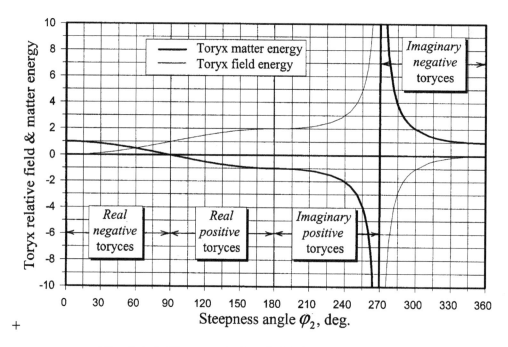

Fig. 7.2.3. Toryx relative field & matter energy.

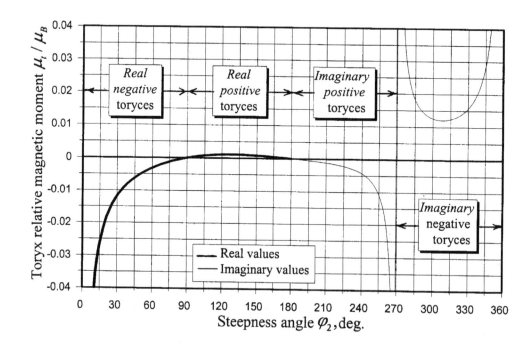

Fig. 7.2.4. Toryx relative magnetic moment.

The toryx relative magnetic moment in respect to the nuclear magneton μ_t / μ_N is related to the toryx relative magnetic moment in respect to the Bohr magneton μ_t / μ_B by the equation:

$$\frac{\mu_t}{\mu_N} = \frac{\mu_t}{\mu_B} \frac{m_p}{m_e}$$

(7.2-1)

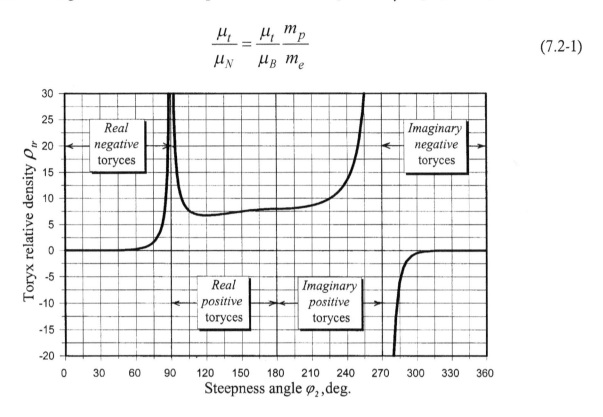

Fig. 7.2.5. Toryx density.

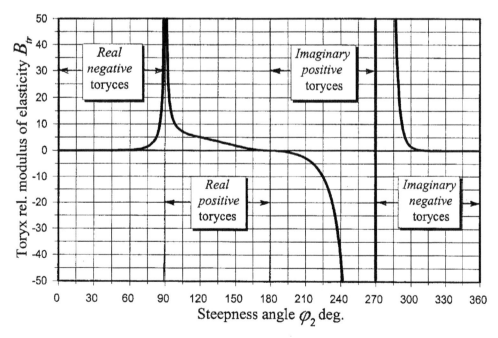

Fig. 7.2.6. Toryx relative bulk modulus of elasticity.

7.3 Toryx Relativistic Equations

Toryx relativistic equations are shown in Table 7.3.

Table 7.3. Toryx relativistic equations.

Relative charge	$\dfrac{e_t}{e_0} = \sqrt{1-\beta_1^2}$	(7.3-1)
Relative gravitational mass	$\dfrac{m_{tg}}{m_e} = Q_q\left|\sqrt{1-\beta_1^2}\right|$	(7.3-2)

- Both masses and charges of toryces are dependent on velocities of their leading strings.
- In real toryces the absolute values of their relative charges and masses decrease with an increase of the relative spiral velocities of their leading strings β_1.
- In imaginary toryces the absolute values of their relative charges and masses increase as the relative spiral velocities of their leading string β_1 increase.

7.4 Relationships between toryx constant spacetime and physical parameters

Table 7.4 shows relationships between toryx constant spacetime and physical parameters.

Table 7.4. Relationships between toryx constant spacetime and physical parameters.

Toryx spacetime Parameters	Equations	
Eye radius	$r_0 = \dfrac{r_e}{2} = \dfrac{e_0^2}{8\pi\varepsilon_0 m_e c^2}$	(7.4-1)
Base frequency	$f_0 = \dfrac{c}{2\pi r_0} = \dfrac{4\varepsilon_0 m_e c^3}{e_0^2}$	(7.4-2)
Base period	$T_0 = \dfrac{2\pi r_0}{c} = \dfrac{e_0^2}{4\varepsilon_0 m_e c^3}$	(7.4-3)

MAIN SUMMARY

*What we consider nowadays as physical properties (mass, charge, fields, waves, etc.) are merely the manifestations of the spiral spacetimes. The theory describes the properties of entities of the Multiverse by using both the **objective spacetime units** that can be discovered by the inhabitants of advanced civilizations and the **subjective physical units** invented by human being.*

Notes:

8. FORMATION OF ELEMENTARY MATTER PARTICLES

8.1 Quantum Vacuum, Virtual Particles & Toryces

Toryces of a microworld can be of two kinds: virtual *toryces* and *spacetime toryces*.

The virtual toryces are incomplete spacetime toryces with random degrees of spacetime freedom. They reside in the *quantum vacuum* expanding from unity (± 1) towards both *infinity* $(\pm \infty)$ and *infinility* $(\pm 0 = \pm 1/\pm \infty)$ as shown in Fig. 8.1.1. Their spontaneous short-lived appearances are governed by the *Toryx uncertainty principle* shown in Exhibit 8.1.1.

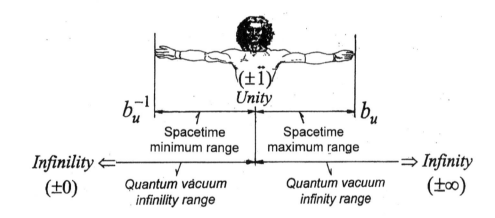

Figure 8.1.1. The ranges occupied by virtual and spacetime toryces.

Exhibit 8.1.1. Toryx Uncertainty Principle.

The product of the *toryx intensity* I and the cycle time of the toryx leading string T_1 must be either equal or smaller than the Planck constant h over 2π as given by the equation:

$$I T_1 \leq \frac{h}{2\pi} = \frac{e_0^2}{4\pi c \varepsilon_0 \alpha_s} \tag{8.1-1}$$

Equation (8.1-1) reduces to the forms:

$$\text{Real toryces} \qquad \frac{b_1 \sqrt{2b_1 - 1}}{b_1 - 1} \leq \frac{1}{Q_q \pi \alpha_s} \tag{8.1-2a}$$

$$\text{Imaginary toryces} \quad \frac{b_1\sqrt{2b_1-1}}{b_1-1} \leq \frac{i}{Q_q\pi\alpha_s} \qquad (8.1\text{-}2b)$$

Based on the Toryx uncertainty principle, the relative radii of leading strings b_1 of virtual toryces are within the ranges shown below.

$$\text{Real toryces} \quad 1.0240338955 \leq b_1 \leq 949.84653950 \qquad (8.1\text{-}3a)$$

$$\text{Imaginary toryces} \quad -1.022925306 \geq b_1 \geq -952.84653950 \qquad (8.1\text{-}3b)$$

The spacetime toryces are created from rare virtual toryces complying with ***toryx genetic codes***. Consequently, they have certain degrees of spacetime freedom and appear in the forms shown in Figs. 1.1.1 – 1.1.3. The spacetime toryces reside within the spacetime ranges extending from unity (± 1) towards both the ultimate maximum value b_u and the ultimate minimum value $b_u^{-1} = 1/b_u$ as shown in Fig. 8.1.1. They exist at certain excitation, oscillation and spacetime quantum states and sustain their existence by cyclic absorption and release of spacetime.

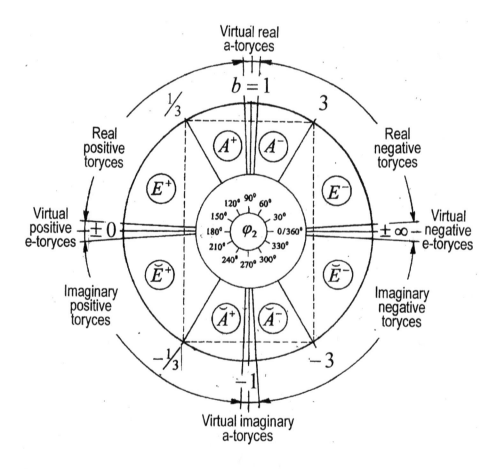

Figure 8.1.2. Ranges of existence of real and imaginary spacetime toryces.

Exhibit 8.1.2 shows the ultimate maximum relative radius b_{1u} of leading strings of real negative toryx in respect to the maximum toryx oscillation factor Q_{qm} defined by Eq. (6.3-2).

Exhibit 8.1.2. The ultimate maximum relative radius b_{1u} of leading strings of real negative toryx in respect to the maximum toryx oscillation factor Q_{qm} .

$$b_{1u} = Q_{qm} = 2.637728 \times 10^{11} \qquad (8.1\text{-}4)$$

Figure 8.1.2 and Table 8.1 show ranges of the relative radii b of spherical boundaries of virtual toryces located beyond the range of the spacetime toryces derived from equations shown in Table 6.2.1 and Eqs. (1.6-13), (6.2-1) and (8.1-4).

Table 8.1. The ranges of the relative radii b of spherical boundaries of virtual toryces located beyond the range of the spacetime toryces.

Virtual toryces	Steepness angle	Ranges of the relative radii b	
		From	To
Real negative e-toryces	$\rightarrow 0/360^0$	b_u	$\rightarrow +\infty$
Imaginary negative e-toryces		$-b_u$	$\rightarrow -\infty$
Real positive e-toryces	$\rightarrow 0/180^0$	b_u^{-1}	$\rightarrow +0$
Imaginary positive e-toryces		$-b_u^{-1}$	$\rightarrow -0$
Real negative a-toryces	$\rightarrow 0/90^0$	$(b_u+1)/(b_u-1)$	$\rightarrow 1$
Real positive a-toryces		$(b_u-1)/(b_u+1)$	$\rightarrow 1$
Imaginary positive a-toryces	$\rightarrow 0/270^0$	$-(b_u+1)/(b_u-1)$	$\rightarrow -1$
Imaginary negative a-toryces		$-(b_u-1)/(b_u+1)$	$\rightarrow -1$

8.2 Formation of Mutually- & Self-polarized Elementary Matter Particles

Mutually-polarized elementary matter particles - As shown in Fig 8.2.1, four kinds of mutually-polarized elementary particles (trons) are formed by the unification of adjacent toryces: the reality-polarized ***electrons*** and ***positrons,*** and the vorticity-polarized ***ethertrons*** and ***singulatrons***.

Self-polarized elementary matter particles - Figure 3.4.2 shows the ranges of the toryx relative radii of leading string b_1 within which the self-polarized and mutually-polarized toryces may exist. Figure 8.2.2 shows a formation of two self-polarized golden trons by unification of adjacent self-polarized toryces.

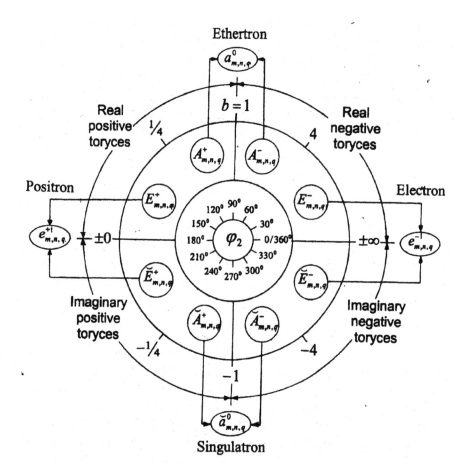

Figure 8.2.1. Formation of four mutually-polarized elementary matter particles (trons).

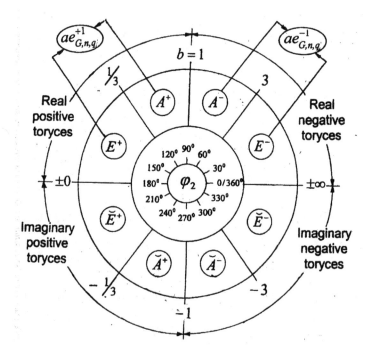

Figure 8.2.2. Formation of two self-polarized golden trons.

8.3 Rules of Formation of Stable Elementary Matter Particles

Reality and vorticity-polarized trons sustain their existence by periodic absorption and release of spacetime (see Exhibits 8.3.1 and 8.3.2).

Exhibit 8.3.1. Tron spacetime quantity absorbed or released by toryces.

The **spacetime quantity** S_t and $\breve{S}_t$ absorbed or released by the real and imaginary toryces are assumed to be proportional to the product of their vorticities V and the square of realities R as given by the equations:

$$S_t = Q_q V R^2 = -\frac{Q_q (b_1 - 1)(2b_1 - 1)}{b_1}$$

$$\breve{S}_t = \breve{Q}_q \breve{V} \breve{R}^2 = -\frac{\breve{Q}_q (\breve{b}_1 - 1)(2\breve{b}_1 - 1)}{\breve{b}_1} \qquad (8.3\text{-}1)$$

Exhibit 8.3.2. Tron spacetime quantity conservation law.

According to the tron spacetime conservation law, the spacetime quantity S of a tron made up of N real and $\breve{N}$ imaginary positive and negative toryces with the respective toryx spacetime quantities S_t and $\breve{S}_t$ must approach the minimum toryx limits b_u^{-1} as given by the equation:

$$S = (S_t N + \breve{S}_t \breve{N}) \rightarrow \pm b_u^{-1} \qquad (8.3\text{-}2)$$

The following rules are applied to the calculations of parameters of mutually-polarized and self-polarized trons:

Mutually-polarized trons are made up of either vorticity-polarized or reality-polarized toryces, alternatively absorbing and releasing spacetime to sustain their existence. Consequently, the total values X_Σ of parameters of the mutually-polarized trons are equal to the arithmetic average sums of the parameters X_1 and $\breve{X}_2$ of their constituent toryces.

Self-polarized trons are made up of either real or imaginary self-polarized toryces coexisting *concurrently* inside their trons thanks to the ability of their constituent toryces to sustain their existence independently by absorbing and releasing spacetime. Consequently, the tron parameters are equal to the arithmetic sums of parameters X_1 and X_2 of their respective constituent toryces.

Two requirements must be met to form stable tron assemblies:

1. In a stable tron made up of N real and $\breve{N}$ imaginary toryces, the **stable tron reality ratio** T is defined by the equation:

$$T = \frac{N}{\breve{N}} = \left(\frac{b_1}{b_1 - 1}\right)^2 = \left(\frac{b+1}{b-1}\right)^2 \tag{8.3-3}$$

2. The trons must be at the lowest oscillation quantum state $q = 0$.

MAIN SUMMARY

*Nature resides in two regions called **quantum vacuum** and **spacetime.***

Quantum vacuum region** extends from **infinity $(\pm\infty)$ *to **infinility*** (± 0)*. It is filled with short-lived **virtual toryces**. Their spontaneous appearances are governed by the **Toryx uncertainty principle** – see Exhibit (8.1.1).*

***Spacetime region** extends from the maximum value* b_u *to the minimum value* b_u^{-1}*. It is occupied by the spacetime toryces created from rare virtual toryces which degrees of freedom are limited by the **toryx genetic codes**. Consequently, they are able to exist at certain excitation, oscillation and spacetime quantum states; they are also able to sustain their existence by cyclic absorption and release of spacetime.*

***Elementary matter particles (trons)** are made up of either **mutually-polarized** or **self-polarized** **toryces** – see Fig. (8.2.1).*

Mutually-polarized elementary matter particles *include:*

- **Electrons** *made up of reality-polarized negative toryces*
- **Positrons** *made up of reality-polarized positive toryces*
- **Ethertrons** – *made up of vorticity-polarized real negative and positive toryces*
- **Singulatrons** – *made up of vorticity-polarized imaginary negative and positive toryces.*

*The constituent toryces of mutually-polarized elementary matter particles **alternatively** absorb and release spacetime to sustain their existence. Consequently, the total parameters of the mutually-polarized trons are equal to the arithmetic average sums of parameters of their constituent toryces.*

Self-polarized elementary matter particles *Fig. (3.4.2) are made up of self-polarized real toryces coexisting **concurrently** inside their trons thanks to the ability of each of their constituent toryces to absorb and release spacetime independently to sustain their existence. Consequently, the tron parameters are equal to the arithmetic sums of parameters of their respective constituent toryces.*

*Stable trons follow the **Tron spacetime quantity conservation law** assuring a balanced absorption and release of spacetime by their constituent toryces – see Exhibits (8.3.1) and (8.3.2).*

<u>Notes:</u>

9. EXAMPLES OF MATTER PARTICLES

CONTENTS

.

9.1 *Reality-polarized basic lambda electron* $e_{0,0,0}^{-1}$ is made up of two real negative basic harmonic toryces $E_{0,0,0}^{-\frac{1}{2}}$ and one half of the imaginary negative basic harmonic toryx $\breve{E}_{0,0,0}^{-2}$ as shown by the equation:

$$e_{0,0,0}^{-1} = 2E_{0,0,0}^{-\frac{1}{2}} + \tfrac{1}{2}\breve{E}_{0,0,0}^{-2} \qquad (9.1\text{-}1)$$

Figure 9.1 and Table 9.1 show cross-section, dimensions, compositions and properties of the reality-polarized basic lambda electron $e_{0,0,0}^{-1}$.

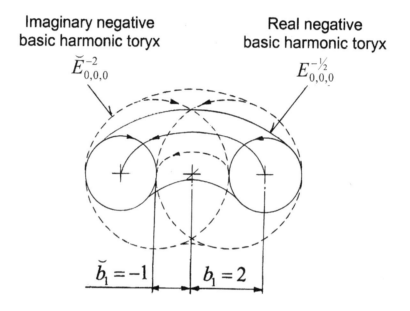

Figure 9.1. Cross-section and dimensions of the reality-polarized basic lambda electron $e_{0,0,0}^{-1}$ and its constituent toryces.

Table 9.1. Composition and properties of the reality-polarized basic lambda electron $e_{0,0,0}^{-1}$ and its constituent toryces.

Toryx	b_1	$\beta_1 = \beta_{2t}$	w_2	e_t/e_0	μ_t/μ_N	m_{tg}/m_e	S	N
$E_{0,0,0}^{-\frac{1}{2}}$	2.0	0.866025	1.154701	-0.50	- 5.80196032	0.500	-1.5	2
$\breve{E}_{0,0,0}^{-2}$	-1.0	-1.732051i	0.577350i	-2.00	- 23.20784127	2.000	+6.0	0.5
Reality-polarized basic lambda electron $e_{0,0,0}^{-1}$	**-1.00**	**-11.60392064**	**1.000**	**0.00**	**1**			

9.2 *Reality-polarized basic lambda positron* $e_{0,0,0}^{+1}$ is made up of two real positive harmonic lambda toryces $E_{0,0,0}^{+\frac{1}{2}}$ and one half of the imaginary positive harmonic lambda toryx $\breve{E}_{0,0,0}^{+2}$ as shown by the equation:

$$e_{0,0,0}^{+1} = 2E_{0,0,0}^{+\frac{1}{2}} + \frac{1}{2}\breve{E}_{0,0,0}^{+2} \qquad (9.2\text{-}1)$$

Figure 9.2 and Table 9.2 show cross-section, dimensions, compositions and properties of the reality-polarized basic lambda positron $e_{0,0,0}^{+1}$.

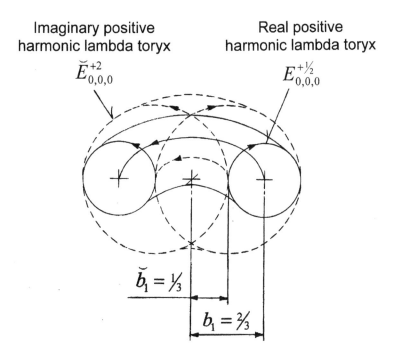

Figure 9.2. Cross-section and dimensions of the reality-polarized basic lambda positron $e_{0,0,0}^{+1}$ and its constituent toryces.

Table 9.2. Composition and properties of the reality-polarized basic lambda positron $e_{0,0,0}^{+1}$ and its constituent toryces.

Toryx	b_1	$\beta_1 = \beta_{2t}$	w_2	e_t/e_0	μ_t/μ_N	m_{tg}/m_e	S	N
$E_{0,0,0}^{+\frac{1}{2}}$	$\frac{2}{3}$	0.866025	1.154701	+ 0.50	+1.93398677	0.500	$+\frac{1}{6}$	2
$\breve{E}_{0,0,0}^{+2}$	$\frac{1}{3}$	- 1.732051i	0.577350i	+ 2.00	+9.63594709	2.000	$-\frac{2}{3}$	0.5
Reality-polarized basic lambda positron $e_{0,0,0}^{+1}$				+1.00	+3.86797355	1.000	0.00	1

9.3 Vorticity-polarized basic lambda ethertron $a_{0,0,0}^0$ is made up of one real negative basic harmonic toryx $A_{0,0,0}^{-\frac{1}{2}}$ and one real positive basic harmonic toryx $A_{0,0,0}^{+\frac{1}{2}}$ as shown by the equation:

$$a_{0,0,0}^0 = A_{0,0,0}^{-\frac{1}{2}} + A_{0,0,0}^{+\frac{1}{2}} \qquad (9.3\text{-}1)$$

Figure 9.3 and Table 9.3 show cross-section, dimensions, compositions and properties of the basic lambda ethertron $a_{0,0,0}^0$.

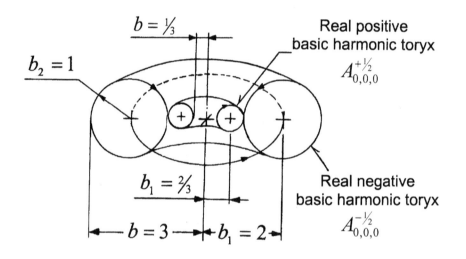

Figure 9.3. Cross-section and dimensions of the vorticity-polarized basic lambda ethertron $a_{0,0,0}^0$ and its constituent toryces.

Table 9.3. Composition and properties of the vorticity-polarized basic lambda ethertron $a_{0,0,0}^0$ and its constituent toryces.

Toryx	b_1	$\beta_1 = \beta_{2t}$	w_2	e_t/e_0	μ_t/μ_N	m_{tg}/m_e	S	N
$A_{0,0,0}^{-\frac{1}{2}}$	2.0	0.866025	1.154701	-0.50	-5.80196032	0.500	-1.5	1
$A_{0,0,0}^{+\frac{1}{2}}$	$\frac{2}{3}$	0.866025	1.154701	+0.50	+1.93398677	0.500	$+\frac{1}{6}$	1
Vorticity-polarized basic lambda ethertron $a_{0,0,0}^0$				**0.00**	**-1.93398677**	**0.500**	$-\frac{2}{3}$	**1**

9.4 *Vorticity-polarized basic lambda singulatron* $\breve{a}^0_{0,0,0}$ is made up of one harmonic imaginary negative basic harmonic toryx $\breve{A}^{-2}_{0,0,0}$ and one imaginary positive basic harmonic toryx $\breve{A}^{+2}_{0,0,0}$ as shown by the equation:

$$\breve{a}^0_{0,0,0} = \breve{A}^{-2}_{0,0,0} + \breve{A}^{+2}_{0,0,0} \tag{9.4-1}$$

Figure 9.4 and Table 9.4 show cross-section, dimensions, compositions and properties of the vorticity-polarized basic lambda singulatron $\breve{a}^0_{0,0,0}$.

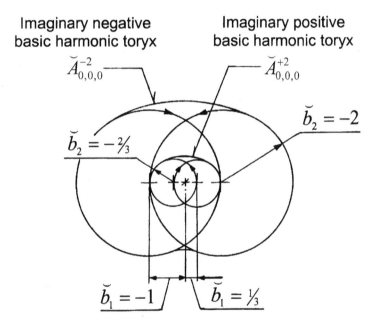

Figure 9.4 Cross-section and dimensions of the vorticity-polarized basic lambda singulatron $\breve{a}^0_{0,0,0}$ and its constituent toryces.

Table 9.4. Composition and properties of the vorticity-polarized basic lambda singulatron $\breve{a}^0_{0,0,0}$ and its constituent toryces.

Toryx	b_1	$\beta_1 = \beta_{2t}$	w_2	e_t/e_0	μ_t/μ_N	m_{tg}/m_e	S	N
$\breve{A}^{-2}_{0,0,0}$	-1.0	- 1.732051i	- 0.577350i	-2.0	-23.20784127	2.000	+6.0	1
$\breve{A}^{+2}_{0,0,0}$	$\frac{1}{3}$	1.732051i	0.577350i	+2.0	+7.73594709	2.000	$-\frac{2}{3}$	1
Vorticity-polarized basic lambda singulatron $\breve{a}^0_{0,0,0}$				**0.00**	**-7.73594709**	**2.000**	$+\frac{8}{3}$	**1**

9.5 Self-polarized basic harmonic electron $ae_{H,0,0}^{-1}$ is made up of one self-polarized real negative harmonic toryx $A_{H,0,0}^{-1/2}$ and one self-polarized real negative harmonic toryx $E_{H,0,0}^{-1/2}$ as given by the equation:

$$ae_{H,0,0}^{-1} = A_{H,0,0}^{-1/2} + E_{H,0,0}^{-1/2} \qquad (9.5\text{-}1)$$

Figure 9.5 and Table 9.5 show cross-section, dimensions, compositions and properties of the self-polarized basic harmonic electron $ae_{H,0,0}^{-1}$.

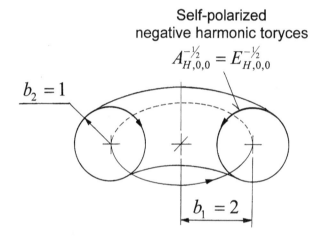

Figure 9.5. Cross-section and dimensions of the self-polarized basic harmonic electron $ae_{H,0,0}^{-1}$ and its constituent toryces.

Table 9.5. Composition and properties of the self-polarized basic harmonic electron $ae_{H,0,0}^{-1}$ and its constituent toryces.

Toryx	b_1	$\beta_1 = \beta_{2t}$	w_2	e_t/e_0	μ_t/μ_N	m_{tg}/m_e	N
$A_{H,0,0}^{-1/2}$	2.0	0.866025	1.154701	-0.50	-5.80196032	0.500	1
$E_{H,0,0}^{-1/2}$	2.0	0.866025	1.154701	-0.50	-5.80196032	0.500	1
Self-polarized basic harmonic electron $ae_{H,0,0}^{-1}$				**-1.00**	**-11.60392064**	**1.000**	**1**

9.6 *Self-polarized basic harmonic positron* $ae^{+1}_{H,0,0}$ is made up of one self-polarized real positive harmonic toryx $A^{+1/2}_{H,0,0}$ and one self-polarized real positive harmonic toryx $E^{+1/2}_{H,0,0}$ as the equation:

$$ae^{+1}_{H,0,0} = A^{+1/2}_{H,0,0} + E^{+1/2}_{H,0,0} \qquad (9.6\text{-}1)$$

Figure 9.6 and Table 9.6 show cross-section, dimensions, compositions and properties of the self-polarized basic harmonic positron $ae^{+1}_{H,0,0}$.

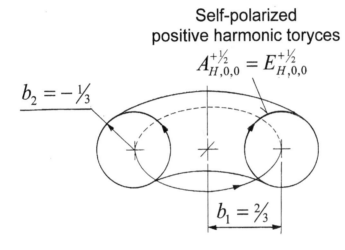

Figure 9.6. Cross-section and dimensions of the self-polarized basic harmonic positron $ae^{+1}_{H,0,0}$ and its constituent toryces.

Table 9.6. Composition and properties of the self-polarized basic harmonic positron $ae^{+1}_{H,0,0}$ and its constituent toryces.

Toryx	b_1	$\beta_1 = \beta_{2t}$	w_2	e_t/e_0	μ_t / μ_N	m_{tg} / m_e	N
$A^{+1/2}_{H,0,0}$	$2/3$	0.866025	1.154701	+0.50	+1.93398677	0.500	1
$E^{+1/2}_{H,0,0}$	$2/3$	0.866025	1.154701	+0.50	+1.93398677	0.500	1
Self-polarized basic harmonic positron $ae^{+1}_{H,0,0}$				**+1.00**	**+ 3.86797355**	**1.000**	**1**

9.7 *Self-polarized excited harmonic electron* $ae_{H,1,0}^{-1}$ is made up of one self-polarized real negative excited harmonic toryx $A_{H,1,0}^{-1/3}$ and one self-polarized real negative excited harmonic toryx $E_{H,1,0}^{-2/3}$ as given by the equation:

$$ae_{H,1,0}^{-1} = A_{H,1,0}^{-1/3} + E_{H,1,0}^{-2/3} \qquad (9.7\text{-}1)$$

Figure 9.7 and Table 9.7 show cross-section, dimensions, compositions and properties of the self-polarized excited harmonic electron $ae_{H,1,0}^{-1}$.

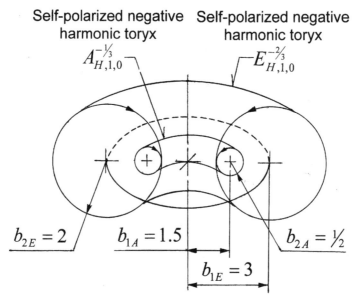

Figure 9.7. Cross-section and dimensions of the self-polarized excited harmonic electron $ae_{H,1,0}^{-1}$ and its constituent toryces.

Table 9.7. Composition and properties of the self-polarized excited harmonic electron $ae_{H,1,0}^{-1}$ and its constituent toryces.

Toryx	b_1	$\beta_1 = \beta_{2t}$	w_2	e_t/e_0	μ_t / μ_N	m_{tg}/m_e	N
$A_{H,1,0}^{-1/3}$	1.5	0.942809	$1.060660i$	$-1/3$	-3.15818717	$1/3$	1
$E_{H,1,0}^{-2/3}$	3.0	0.745356	1.341641	$-2/3$	-9.98706475	$2/3$	1
Self-polarized excited harmonic electron $ae_{H,1,0}^{-1}$	**-1.00**				**-13.14525192**	**1.000**	**1**

9.8 *Self-polarized excited harmonic positron* $ae_{H,1,0}^{+1}$ is made up of one self-polarized real positive excited harmonic toryx $A_{H,1,0}^{+1/3}$ and one self-polarized real positive excited harmonic toryx $E_{H,1,0}^{+2/3}$ as given by the equation:

$$ae_{H,1,0}^{+1} = A_{H,1,0}^{+1/3} + E_{H,1,0}^{+2/3} \tag{9.8-1}$$

Figure 9.8 and Table 9.8 show cross-section, dimensions, compositions and properties of the self-polarized excited harmonic positron $ae_{H,1,0}^{+1}$.

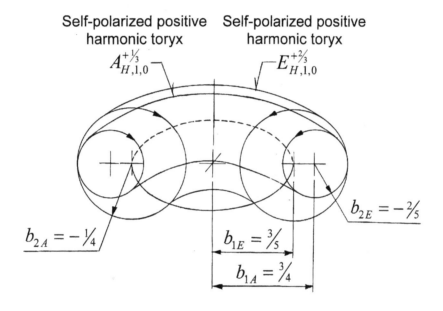

Figure 9.8. Cross-section and dimensions of the self-polarized excited harmonic positron $ae_{H,1,0}^{+1}$ and its constituent toryces.

Table 9.8. Composition and properties of the self-polarized excited harmonic positron $ae_{H,1,0}^{+1}$ and its constituent toryces.

Toryx	b_1	$\beta_1 = \beta_{2t}$	w_2	e_t/e_0	μ_t / μ_N	m_{tg} / m_e	N
$A_{H,1,0}^{+1/3}$	$3/4$	0.942809	1.060660	$+1/3$	+1.57909359	$1/3$	1
$E_{H,1,0}^{+2/3}$	$3/5$	0.745356	1.341641	$+2/3$	+1.99741295	$2/3$	1
Self-polarized excited harmonic positron $ae_{H,1,0}^{+1}$				**+1.00**	**+3.57650654**	**1.000**	**1**

9.9 Self-polarized basic golden electron $ae_{G,0,0}^{-1}$ of all spacetime levels of the Multiverse is made up of one self-polarized negative golden toryx $A_{G,0,0}^{-1/2}$ and one self-polarized negative golden toryx $E_{G,0,0}^{-1/2}$ as given by the equation:

$$ae_{G,0,0}^{-1} = A_{G,0,0}^{-1/2} + E_{G,0,0}^{-1/2} \tag{9.9-1}$$

Figure 9.9 and Table 9.9 show cross-section, dimensions, compositions and properties of the self-polarized basic golden electron $ae_{G,0,0}^{-1}$.

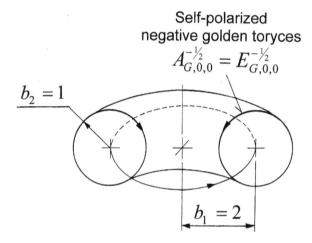

Figure 9.9. Cross-section and dimensions of the self-polarized basic golden electron $ae_{G,0,0}^{-1}$ and its constituent toryces.

Table 9.9. Composition and properties of the self-polarized basic golden electron $ae_{G,0,0}^{-1}$ and its constituent toryces.

Toryx	b_1	$\beta_1 = \beta_{2t}$	w_2	e_t/e_0	μ_t / μ_N	m_{tg} / m_e	N
$A_{G,0,0}^{-1/2}$	2.0	0.866025	1.154701	-0.50	-5.80196032	0.500	1
$E_{G,0,0}^{-1/2}$	2.0	0.866025	1.154701	-0.50	-5.80196032	0.500	1
Self-polarized basic golden electron $ae_{G,0,0}^{-1}$				**-1.00**	**-11.60392064**	**1.000**	**1**

9.10 *Self-polarized basic golden positron* $ae_{G,0,0}^{+1}$ of all spacetime levels of the Multiverse is made up of one self-polarized positive golden toryx $A_{G,0,0}^{+1/2}$ and one self-polarized basic positive golden toryx $E_{G,0,0}^{+1/2}$ as given by the equation:

$$ae_{G,0,0}^{+1} = A_{G,0,0}^{+1/2} + E_{G,0,0}^{+1/2} \qquad (9.10\text{-}1)$$

Figure 9.10 and Table 9.10 show cross-section, dimensions, compositions and properties of the self-polarized basic golden positron $ae_{G,0,0}^{+1}$.

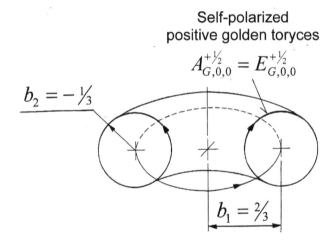

Figure 9.10. Cross-section and dimensions of the self-polarized basic golden positron $ae_{G,0,0}^{+1}$ and its constituent toryces.

Table 9.10. Composition and properties of the self-polarized basic golden positron $ae_{G,0,0}^{+1}$ and its constituent toryces.

Toryx	b_1	$\beta_1 = \beta_{2t}$	w_2	e_t/e_0	μ_t/μ_N	m_{tg}/m_e	N
$A_{G,0,0}^{+1/2}$	$2/3$	0.866025	1.154701	+0.50	+1.93398677	0.500	1
$E_{G,0,0}^{+1/2}$	$2/3$	0.866025	1.154701	+0.50	+1.93398677	0.500	1
Self-polarized basic golden positron $ae_{G,0,0}^{+1}$				+1.00	+ 3.86797355	1.000	1

9.11 *Self-polarized excited golden electron* $ae_{G,1,0}^{-1}$ of all spacetime levels of the Multiverse is made up of one self-polarized real negative excited golden toryx $A_{G,1,0}^{-\phi^{-2}}$ and one self-polarized real negative excited golden toryx $E_{G,1,0}^{-\phi^{-1}}$ as by the equation:

$$ae_{G,1,0}^{-1} = A_{G,1,0}^{-\phi^{-2}} + E_{G,1,0}^{-\phi^{-1}} \tag{9.11-1}$$

Figure 9.11 and Table 9.11 show cross-section, dimensions, compositions and properties of the self-polarized excited golden electron $ae_{G,1,0}^{-1}$.

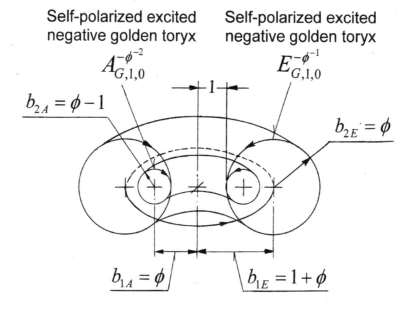

Figure 9.9.25. Cross-section and dimensions of the self-polarized excited golden electron $ae_{G,1,0}^{-1}$ and its constituent toryces.

Table 9.9.25. Composition and properties of the self-polarized excited golden electron $ae_{G,1,0}^{-1}$ and its constituent toryces.

Toryx	b_1	$\beta_1 = \beta_{2t}$	w_2	e_t/e_0	μ_t/μ_N	m_{tg}/m_e	N
$A_{G,1,0}^{-\phi^{-2}}$	ϕ	0.924176	1.082045	$-\phi^{-2}$	- 3.8265848	ϕ^{-2}	1
$E_{G,1,0}^{-\phi^{-1}}$	$1+\phi$	0.786151	1.272020	$-\phi^{-1}$	- 8.5219296	ϕ^{-1}	1
Self-polarized excited golden electron $ae_{G,1,0}^{-1}$				**-1.00**	**-12.3485144**	**1.000**	**1**

9.12 *Self-polarized excited golden positron* $ae_{G,1,0}^{+1}$ of all spacetime levels of the Multiverse is made up of one self-polarized real positive excited golden toryx $A_{G,1,0}^{+\phi^{-2}}$ and one self-polarized real positive excited golden toryx $E_{G,1,0}^{+\phi^{-1}}$ as given by the equation:

$$ae_{G,1,0}^{+1} = A_{G,1,0}^{+\phi^{-2}} + E_{G,1,0}^{+\phi^{-1}} \qquad (9.12\text{-}1)$$

Figure 9.12 and Table 9.12 show cross-section, dimensions, compositions and properties of the self-polarized excited golden electron $ae_{G,1,0}^{-1}$.

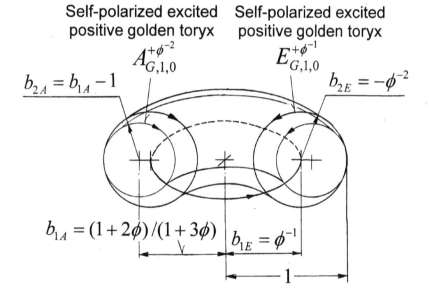

Figure 9.12. Cross-section and dimensions of the self-polarized excited golden positron $ae_{G,1,0}^{+1}$ and its constituent toryces.

Table 9.12. Composition and properties of the self-polarized excited golden positron $ae_{G,1,0}^{+1}$ and its constituent toryces.

Toryx	b_1	$\beta_1 = \beta_{2t}$	w_2	e_t/e_0	μ_t/μ_N	m_{tg}/m_e	N
$A_{G,1,0}^{+\phi^{-2}}$	$\frac{1+2\phi}{1+3\phi}$	0.924176	1.082045	$+\phi^{-2}$	+1.7113008	ϕ^{-2}	1
$E_{G,1,0}^{+\phi^{-1}}$	ϕ^{-1}	0.786151	1.272020	$+\phi^{-1}$	+2.0117547	ϕ^{-1}	1
Self-polarized excited golden positron $ae_{G,1,0}^{+1}$				+1.00	+3.7230554	1.000	1

9.13 *Reality-polarized excited lambda electron* $e_{m,n,q}^{-1}$ is made up of one excited real negative lambda toryx $E_{m,n,q}^{-}$ and one excited imaginary negative lambda toryx $\breve{E}_{m,n,q}^{-}$ as shown by the equation:

$$e_{m,n,q}^{-1} = E_{m,n,q}^{-} + \breve{E}_{m,n,q}^{-} \tag{9.13-1}$$

Figure 9.13 and Table 9.13 show cross-section, dimensions, compositions and properties of the reality-polarized excited lambda electron $e_{2,1,0}^{-1}$ of ordinary matter.

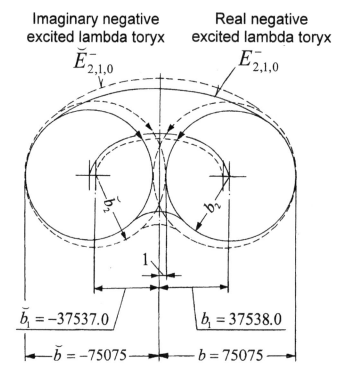

Figure 9.13.   Cross-section and dimensions of the reality-polarized excited lambda electron $e_{2,1,0}^{-1}$ of ordinary matter.

Table 9.13.   Composition and properties of the excited lambda electron $e_{2,1,0}^{-1}$ of ordinary matter.

Toryx	b_1	e_t/e_0	μ_t/μ_N	m_{tg}/m_e	S	N
$E_{2,1,0}^{-}$	37539.0	-0.99997336	-1835.609190	0.99997336	-75075.0	1.00002664
$\breve{E}_{2,1,0}^{-}$	-37537.0	-1.00002664	-1835.706994	1.00002664	+75075.0	0.99997336
Reality-polarized excited lambda electron $e_{2,1,0}^{-1}$		- 1.0000000	-1835.658091	1.00000000	0.000	1

9.14 *Reality-polarized excited lambda positron* $e_{m,n,q}^{+1}$ is made up of one excited real positive lambda toryx $E_{m,n,q}^{+}$ and one excited imaginary positive lambda toryx $\breve{E}_{m,n,q}^{+}$ as shown by the equation:

$$e_{m,n,q}^{+1} = E_{m,n,q}^{+} + \breve{E}_{m,n,q}^{+} \tag{9.14-1}$$

Figure 9.14 and Table 9.14 show cross-section, dimensions, compositions and properties of the reality-polarized excited lambda positron $e_{2,1,0}^{+1}$ of ordinary matter.

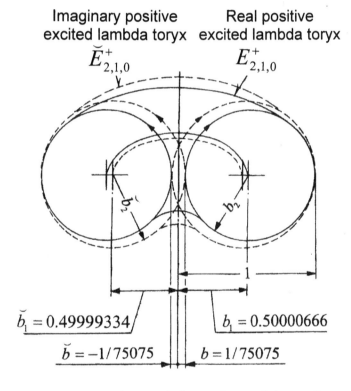

Figure 9.14. Cross-section and dimensions of the reality-polarized excited lambda positron $e_{2,1,0}^{+1}$ of ordinary matter.

Table 9.14. Composition and properties of the excited lambda positron $e_{2,1,0}^{+1}$ of ordinary matter.

Toryx	b_1	e_t/e_0	μ_t/μ_N	m_{tg}/m_e	S	N
$E_{2,1,0}^{+}$	0.50000666	+0.99997336	+0.024450	0.99997336	$+1.332\times10^{-5}$	1.00002664
$\breve{E}_{2,1,0}^{+}$	0.49999334	+1.00002664	+0.024452	1.00002664	-1.332×10^{-5}	0.99997336
Reality-polarized excited lambda positron $e_{2,1,0}^{+1}$		+1.0000000	+0.024451	1.00000000	0.000	1

9.15 *Vorticity-polarized excited lambda ethertron* $a^0_{m,n,q}$ is made up of one excited real negative lambda toryx $A^-_{m,n,q}$ and one excited real positive lambda toryx $A^+_{m,n,q}$ as shown by the equation:

$$a^0_{m,n,q} = A^-_{m,n,q} + A^+_{m,n,q} \qquad (9.15\text{-}1)$$

Figure 9.15 and Table 9.15 show cross-section, dimensions, compositions and properties of the vorticity-polarized excited lambda ethertron $a^0_{1,1,0}$ of ordinary matter.

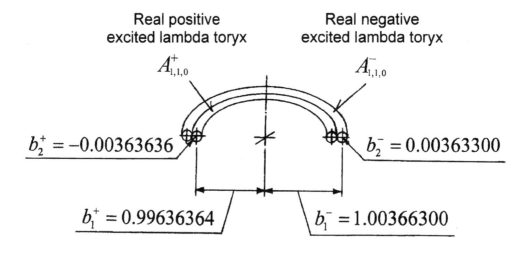

Figure 9.15.   Cross-section and dimensions of the vorticity-polarized excited lambda ethertron $a^0_{1,1,0}$ of ordinary matter.

Table 9.15. Composition and properties of the vorticity-polarized excited lambda ethertron $a^0_{1,1,0}$ of ordinary matter.

Toryx	b_1	e_t/e_0	μ_t/μ_N	m_{tg}/m_e	S	N
$A^-_{1,1,0}$	1.00366300	- 0.00364964	-0.02454023	0.00364964	-0.00367637	1
$A^+_{1,1,0}$	0.99636364	+ 0.00364964	+0.02436175	0.00364964	+0.00362309	1
Vorticity-polarized excited lambda ethertron $a^0_{1,1,0}$	**0.00**	**-0.00008924**	**0.00364964**	**-0.00002664**	**1**

9.16 *Vorticity-polarized excited lambda singulatron* $\breve{\tilde{a}}^0_{m,n,q}$ is made up of one excited imaginary negative lambda toryx $\breve{A}^-_{m,n,q}$ and one excited imaginary positive lambda toryx $\breve{A}^+_{m,n,q}$ as shown by the equation:

$$\breve{\tilde{a}}^0_{m,n,q} = \breve{A}^-_{m,n,q} + \breve{A}^+_{m,n,q} \qquad (9.16\text{-}1)$$

Figure 9.16 and Table 9.16 show cross-section, dimensions, compositions and properties of the vorticity-polarized excited lambda singulatron $\breve{\tilde{a}}^0_{1,1,0}$ of ordinary matter.

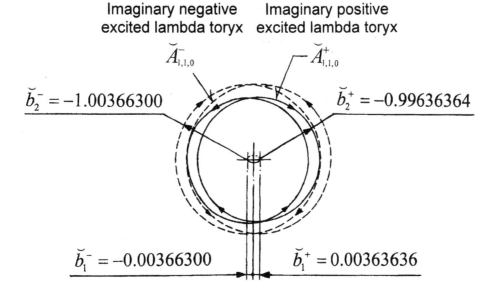

Figure 9.16. Cross-section and dimensions of the vorticity-polarized excited lambda singulatron $\breve{\tilde{a}}^0_{1,1,0}$ of ordinary matter.

Table 9.16. Composition and properties of the vorticity-polarized excited lambda singulatron $\breve{\tilde{a}}^0_{1,1,0}$ of ordinary matter.

Toryx	b_1	e_t/e_0	μ_t/μ_N	m_{tg}/m_e	S	N
$\breve{A}^-_{1,1,0}$	0.00366300	- 274.000000	-1842.382113	274.00	+276.007326	1
$\breve{A}^+_{1,1,0}$	0.00363636	+ 274.000000	+1828.982971	274.00	-272.007273	1
Vorticity-polarized excited lambda singulatron $\breve{\tilde{a}}^0_{1,1,0}$	**0.00**	**- 6.69957132**	**274.00**	**2.000027**	**1**	

9.17 Combined basic lambda ce-tron $ce_{0,0,0}^{0}$ is made up of one harmonic lambda electron $e_{0,0,0}^{-1}$ and one harmonic lambda positron $e_{0,0,0}^{+1}$ as shown by the equation:

$$ce_{0,0,0}^{0} = e_{0,0,0}^{-1} + e_{0,0,0}^{+1} \qquad (9.17\text{-}1)$$

Figure 9.17 and Table 9.17 show cross-section, dimensions, compositions and properties of the combined basic lambda ce-tron $ce_{0,0,0}^{0}$.

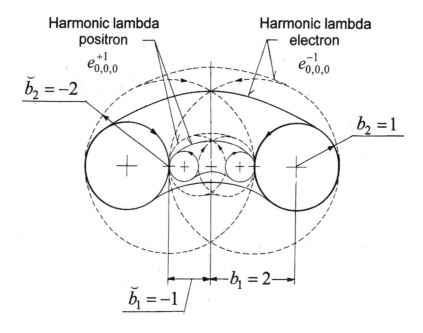

Figure 9.17. Cross-section and dimensions of properties of the combined basic lambda ce-tron $ce_{0,0,0}^{0}$ and its constituent toryces.

Table 9.17. Composition and properties of the combined basic lambda ce-tron $ce_{0,0,0}^{0}$ and its constituent toryces.

Harmonic lambda trons		e_t/e_0	μ_t/μ_N	m_{tg}/m_e	S	N
Electron $e_{0,0,0}^{-1}$	Table 9.1	-1.00	- 11.60392064	1.000	0.00	1
Positron $e_{0,0,0}^{+1}$	Table 9.2	+1.00	+3.86797355	1.000	0.00	1
Combined basic lambda ce-tron $ce_{0,0,0}^{0}$		**0.00**	**-7.73594709**	**2.000**	**0.00**	1

9.18 *Combined basic lambda ca-tron* $ca^0_{0,0,0}$ is made up of one harmonic lambda singulatron $\breve{a}^0_{0,0,0}$ and four harmonic lambda ethertrons $a^0_{0,0,0}$ as shown by the equation:

$$ca^0_{0,0,0} = \breve{a}^0_{0,0,0} + 4a^0_{0,0,0} \qquad (9.18\text{-}1)$$

Figure 9.18 and Table 9.18 show cross-section, dimensions, compositions and properties of the combined basic lambda ca-tron $ca^0_{0,0,0}$.

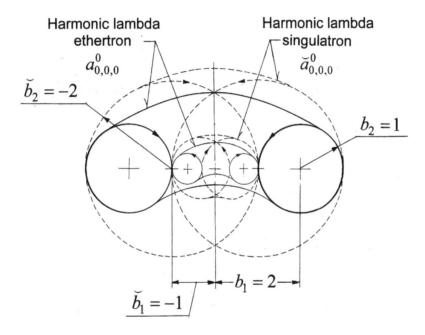

Figure 9.18. Cross-section and dimensions of the combined basic lambda ca-tron $ca^0_{0,0,0}$ and its constituent toryces.

Table 9.18. Composition and properties of the combined basic lambda ca-tron $ca^0_{0,0,0}$ and its constituent toryces.

Harmonic lambda trons		e_t/e_0	μ_t/μ_N	m_{tg}/m_e	S	N
Singulatron $\breve{a}^0_{0,0,0}$	Table 9.4	0.00	-7.73594709	2.000	$+\frac{8}{3}$	1
Ethertron $a^0_{0,0,0}$	Table 9.3	0.00	-1.93398677	0.500	$-\frac{2}{3}$	4
Combined basic lambda ca-tron $ca^0_{0,0,0}$		**0.00**	**-7.73594709**	**2.000**	**0.00**	1

9.19 Combined excited lambda ce-tron $ce^0_{m,n,q}$ is made up of one excited electron $e^{-1}_{m,n,q}$ and one excited positron $e^{+1}_{m,n,q}$ as shown in Fig. 19.19, Table 9.19 and by the equation:

$$ce^0_{m,n,q} = e^{-1}_{m,n,q} + e^{+1}_{m,n,q}$$

(9.19-1)

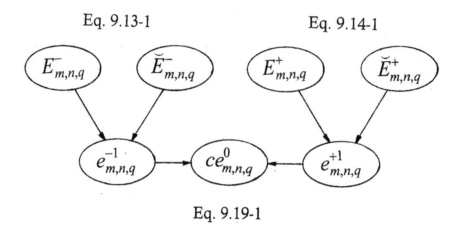

Figure 9.19. Formation of the combined excited lambda ce-tron $ce^0_{m,n,q}$.

Table 9.19. Composition and properties of the combined excited lambda ce-tron $ce^0_{2,1,0}$ of ordinary matter.

Excited lambda trons		e_t/e_0	μ_t/μ_N	m_{tg}/m_e	S	N
Electron $e^{-1}_{2,1,0}$	Table 9.13	- 1.000000	-1835.658091	1.000000	0.0	1
Positron $e^{+1}_{2,1,0}$	Table 9.14	+1.000000	+0.024451	1.000000	0.0	1
Combined excited ce-tron $ce^0_{2,1,0}$		**0.000000**	**-1835.633640**	**2.000000**	**0.0**	**1**

9.20 Combined excited lambda ca-tron $ca^0_{m,n,q}$ is made up of one excited singulatron $\breve{a}^0_{m,n,q}$ and N excited ethertrons $a^0_{m,n,q}$ as shown in Fig. 9.20, Table 9.20 and by the equation:

$$ca^0_{m,n,q} = \breve{a}^0_{m,n,q} + Na^0_{m,n,q} \tag{9.20-1}$$

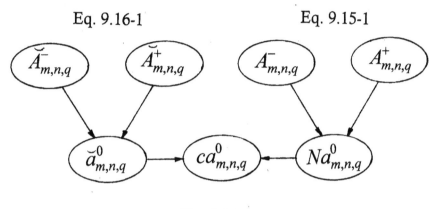

Figure 9.20. Formation of the combined excited lambda ca-tron $ca^0_{m,n,q}$.

Table 9.20. Composition and properties of the combined excited lambda ca-tron $ca^0_{1,1,0}$ of ordinary matter.

Excited lambda trons		e_t/e_0	μ_t/μ_N	m_{tg}/m_e	S	N
Singulatron $\breve{a}^0_{1,1,0}$	Table 9.16	0.00	- 6.69957068	274.000000	+2.000027	1
Ethertron $a^0_{1,1,0}$	Table 9.15	0.00	-0.00008924	0.00364964	-0.000027	75076
Combined excited ca-tron $ca^0_{1,1,0}$		**0.000000**	**- 6.69957132**	**274.000000**	**0.000000**	**1**

9.21 Combined basic and excited golden cae-trons $cae^0_{G,n,q}$ are made up of one golden electron $ae^{-1}_{G,n,q}$ and one golden positron $ae^{+1}_{G,n,q}$ as shown in Fig. 9.21 and by the equation:

$$cae^0_{G,n,q} = ae^{-1}_{G,n,q} + ae^{+1}_{G,n,q} \tag{9.21-1}$$

Figure 9.21 and Table 9.21.1 show the formation, composition and properties of the combined excited golden cae-tron $cae^0_{G,1,0}$ and its constituent toryces.

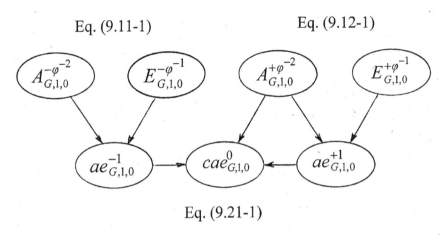

Eq. (9.21-1)

Figure 9.21. Formation of the combined excited golden cae-tron $cae_{G,1,0}^{0}$.

Table 9.21.1. Composition and properties of the combined excited golden cae-tron $cae_{G,1,0}^{0}$ and its constituent toryces.

Excited golden trons		e_t/e_0	μ_t/μ_N	m_{tg}/m_e	S	N
Golden electron $ae_{G,1,0}^{-1}$	Table 9.11	- 1.000000	-12.3485144	1.000000	0.0	1
Golden positron $ae_{G,1,0}^{+1}$	Table 9.12	+1.000000	+3.72305554	1.000000	0.0	1
Combined excited cae-tron $cae_{G,1,0}^{0}$	**0.000000**	**-8.62545898**	**2.000000**	**0.0**	**1**

Table 9.21.2 shows composition and properties of the combined basic golden cae-tron $cae_{G,0,0}^{0}$ and its constituent toryces.

Table 9.21.2. Composition and properties of the combined basic golden cae-tron $cae_{G,0,0}^{0}$ and its constituent toryces.

Excited golden trons		e_t/e_0	μ_t/μ_N	m_{tg}/m_e	S	N
Golden electron $ae_{G,0,0}^{-1}$	Table 9.9	- 1.000000	-11.60392064	1.000000	0.0	1
Golden positron $ae_{G,0,0}^{+1}$	Table 9.10	+1.000000	+3.86797355	1.000000	0.0	1
Combined excited cae-tron $cae_{G,0,0}^{0}$	**0.000000**	**-7.73594709**	**2.000000**	**0.0**	**1**

9.22 Vorticity-polarized oscillated harmonic lambda electron $e_{0,0,1}^{-1}$ is made up of two real negative oscillated harmonic lambda toryx $E_{0,0,1}^{-\frac{1}{2}}$ and one half imaginary negative oscillated harmonic lambda toryx $\breve{E}_{0,0,1}^{-2}$ as shown in Table 9.22 and by the equation:

$$e_{0,0,1}^{-1} = 2E_{0,0,1}^{-\frac{1}{2}} + 0.5\breve{E}_{0,0,1}^{-2} \tag{9.22-1}.$$

Table 9.22. Composition and properties of the vorticity-polarized oscillated lambda electron $e_{0,0,1}^{-1}$ and its constituent toryces.

Trons	b_1	$\beta_1 = \beta_{2t}$	w_2	e_t/e_0	μ_t/μ_N	m_{tg}/m_e	S	N
$E_{0,0,1}^{-\frac{1}{2}}$	2.0	0.866025	1.154701	-0.50	- 1.93398677	1.500	-4.5	2
$\breve{E}_{0,0,1}^{-2}$	-1.0	-1.732051i	0.577350i	-2.00	-7.73594701	6.000	+18.0	0.5
Oscillated lambda electron ↓ $e_{0,0,1}^{-1}$				-1.00	-3.86797355	3.000	0.00	1

9.23 Vorticity-polarized oscillated harmonic lambda ethertron $a_{0,0,1}^{0}$ is made up of one real negative oscillated harmonic lambda toryx $A_{0,0,1}^{-\frac{1}{2}}$ and one real positive oscillated harmonic lambda toryx $A_{0,0,1}^{+\frac{1}{2}}$ as shown in Table 9.23 and by the equation:

$$a_{0,0,1}^{0} = A_{0,0,1}^{-\frac{1}{2}} + A_{0,0,1}^{+\frac{1}{2}} \tag{9.23-1}.$$

Table 9.23. Composition and properties of the vorticity-polarized oscillated lambda ethertron $a_{0,0,1}^{0}$ and its constituent toryces.

Trons	b_1	$\beta_1 = \beta_{2t}$	w_2	e_t/e_0	μ_t/μ_N	m_{tg}/m_e	S	N
$A_{0,0,1}^{-\frac{1}{2}}$	2.0	0.866025	1.154701	-0.50	- 1.93398677	1.500	-4.5	1
$A_{0,0,1}^{+\frac{1}{2}}$	$\frac{2}{3}$	0.866025	1.154701	+ 0.50	+0.64466226	1.500	+0.5	1
Oscillated lambda ethertron ↓ $a_{0,0,1}^{0}$				-1.00	-0.64466226	1.500	-2.00	1

9.24 *Ether* is made up of combined excited lambda ca-trons $ca^0_{m,n,q}$ with each of them made up of one excited singulatron $\breve{a}^0_{m,n,q}$ and N excited ethertrons $a^0_{m,n,q}$ as shown in Figs. 9.20 and 9.24, Table 9.20 and Eq. (9.20-1). In stable trons made up of N real and $\breve{N}$ imaginary toryces, the relationship between the number of real and imaginary toryces are defined by the stable tron reality ratio T defined by the equation (8.3-5).

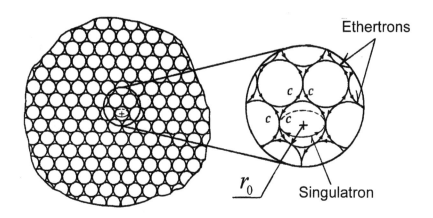

Figure 9.24. Ether.

The ether performs two principal functions:

- The ether serves as a spacetime reservoir used by the vorticity- and reality-polarized toryces for cyclic absorption and release of necessary spacetime to sustain their existence.
- Since the spiral velocities of trailing strings of toryces making up the ethertrons are equal to the velocity of light, the ether provides the velocity reference to all moving entities in all places of the Multiverse.

Table 9.24 shows the stable tron reality ratio T, or the number of ethertrons per one singulatron, in the ethers occupying various spacetime levels of our Multiverse L.

Table 9.24. The stable tron reality ratio T in ethers occupying various spacetime levels of our Multiverse L.

Spacetime levels of our Multiverse					
L0	*L1*	*L2*	*L3*	*L4*	*L5*
The stable tron reality ratio T					
4	7.5076×10^4	1.4091×10^9	2.6447×10^{13}	4.9639×10^{17}	9.3168×10^{21}

Notes:

<u>*Notes:*</u>

10. ETHER, NUCLEONS, LIGHT ATOMS & ISOTOPES

Nucleons (protons and neutrons) are made up of three parts: *nucleon crystal*, *nucleon core* and *nucleon leptons.*

10.1 Nucleon Crystal

The main purpose of the nucleon crystal is to retain excited ethertrons and singulatrons, the constituents of the nucleon core. This is accomplished by locating the ethertrons and singulatrons inside the cavities (Fig. 10.1.1) formed around the center 1 and the vertices 2 through 9 of a nucleon bi-pyramid hexagonal crystal.

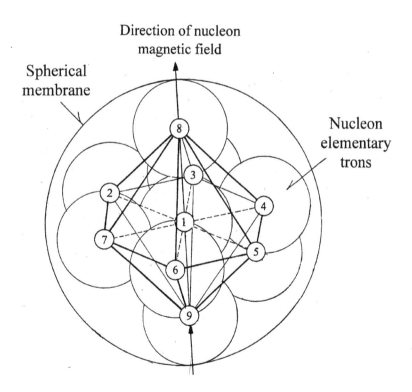

Figure 10.1.1. Isometric view of a nucleon crystal.

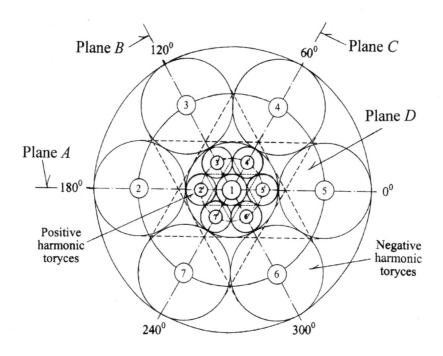

Figure 10.1.2. Cross-section of outer and inner parts of the nucleon crystal $\downarrow nx_{0,0,0}^{0}$.

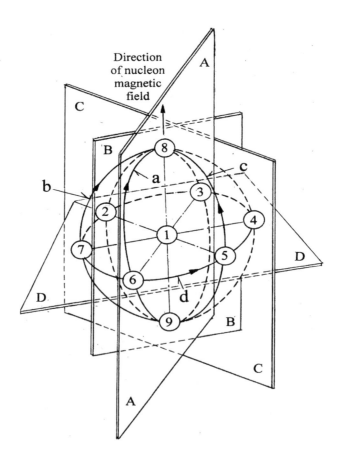

Figure 10.1.3. Isometric view of either outer or inner part of the nucleon crystal $\downarrow nx_{0,0,0}^{0}$.

The nucleon crystal $\downarrow nx_{G,m,n}^0$ is made up of three combined golden cae-trons $cae_{G,n,q}^0$ and one combined golden cae-tron $\downarrow cae_{G,n,q}^0$. Three cae-trons $cae_{G,n,q}^0$ are located in the planes A, B and C, while one cae-tron $\downarrow cae_{G,n,q}^0$ is located in the plane D. The centers of all four ce-trons are located at the nucleon crystal center 1. The directions of the tron magnetic fields are perpendicular to the planes A, B, C and D in which they reside. A contribution of each ce-tron to the total magnetic moment of the nucleon crystal depends on the angle γ between the directions of their magnetic fields and the direction of the nucleon magnetic field shown in Figs. 10.1.1 and 10.1.3. Consequently, because for three golden cae-trons $cae_{G,n,q}^0$ the angle $\gamma = 90^0$, these cae-trons make no contribution to the magnetic moment of the nucleon crystal. At the same time, because for the cae-tron $\downarrow cae_{G,n,q}^0$ the angle $\gamma = 0^0$, this cae-tron makes a full contribution to the magnetic moment of the nucleon crystal.

Based on Tables 9.21.1 and 9.21.2, Tables 10.1.1 and 10.1.2 show compositions and properties of the excited golden nucleon crystal $\downarrow nx_{G,1,0}^0$ and the basic golden nucleon crystal $\downarrow nx_{G,0,0}^0$.

Table 10.1.1. Composition and properties of the excited golden nucleon crystal $\downarrow nx_{G,1,0}^0$.

Combined excited golden cae-trons (Table 9.21.1)	Planes in Figs. 10.1.2, 10.1.3	γ	e_t/e_0	μ/μ_N	m_g/m_e	S	N
Cae-tron $cae_{G,1,0}^0$	$A, B. C$	90^0	0.00	0.00	2.00	0.0	3
Cae-tron $\downarrow cae_{G,1,0}^0$	D	0^0	0.00	-8.62545898	2.00	0.0	1
Excited golden nucleon crystal $\downarrow nx_{G,1,0}^0$	-		**0.00**	**-8.62545898**	**8.00**	**0.0**	**1**

Table 10.1.2. Composition and properties of the basic golden nucleon crystal $\downarrow nx_{G,0,0}^0$.

Combined basic golden cae-trons (Table 9.21.2)	Planes in Figs. 10.1.2, 10.1.3	γ	e_t/e_0	μ/μ_N	m_g/m_e	S	N
Cae-tron $cae_{G,0,0}^0$	$A, B. C$	90^0	0.00	0.00	2.00	0.0	3
Cae-tron $\downarrow cae_{G,0,0}^0$	D	0^0	0.00	-7.73594709	2.00	0.0	1
Basic golden nucleon crystal $\downarrow nx_{G,0,0}^0$	-		**0.00**	**-7.73594709**	**8.00**	**0.0**	**1**

Figure 10.1.2 shows a cross-section of the nucleon crystal. It has two parts, outer and inner. The inner part is located inside the outer part and it is three times smaller than the outer part. Each part has the center 1 and eight vertices 2 through 9. As shown in Figs. 9.9 - 9.12, 10.1.2 and 10.1.3, the vertices of the outer and inner parts are respectively located at the intersections of leading string strings *a*, *b*, *c* and *d* of negative and positive golden toryces forming the combined golden cae-trons.

10.2 Nucleon Core

As shown in Table 10.2, the nucleon core $\uparrow nc_{1,1,0}^{0}$ of the ordinary spacetime level *L1* is made up of three combined excited lambda ca-trons $\downarrow ca_{1,1,0}^{0}$ residing at the nucleon crystal vertices 1, 2, 5 and four combined excited lambda ca-trons $\uparrow ca_{1,1,0}^{0}$ residing at the nucleon crystal vertices 3, 4, 6, 7.

Table 10.2. Composition and properties of the nucleon core $nc_{1,1,0}^{0}$ of the ordinary spacetime level *L1*.

Combined excited trons		Vertices in Figs. 10.1.1 – 10.1.3	Tables	μ/μ_N	m_g/m_e	S	N
Ca-tron $\downarrow ca_{1,1,0}^{0}$	Table 9.20	1, 2, 5	9.15	-6.69957132	274.00	0.0	3
Ca-tron $\uparrow ca_{1,1,0}^{0}$		3, 4, 6, 7	9.16	+6.69957132	274.00	0.0	4
Nucleon core $\uparrow nc_{1,1,0}^{0}$				**+6.69957132**	**1918.00**	**0.0**	**1**

10.3 Nucleons of Ordinary Spacetime Level

Shown below are the structures and properties of neutrons and proton of the ordinary spacetime level *L1*.

Excited oscillated neutron $\downarrow neo_{L1}^{0}$ of the ordinary spacetime level *L1* is made up of one excited golden nucleon crystal $\downarrow nx_{G,1,0}^{0}$, one nucleon core $\uparrow nc_{1,1,0}^{0}$, one basic lambda positron $\uparrow e_{0,0,0}^{+1}$ and one oscillated lambda electron $\downarrow e_{0,0,1}^{-1}$ as shown in Table 10.3.1 and by the equation:

$$\downarrow neo_{L1}^{0} = \downarrow nx_{G,1,0}^{0} + \uparrow nc_{1,1,0}^{0} + \uparrow e_{0,0,0}^{+1} + \downarrow e_{0,0,1}^{-1} \qquad (10.3\text{-}1)$$

Table 10.3.1. Components and properties of the excited oscillated neutron $\downarrow neo^0_{L1}$ of the ordinary spacetime level *L1*.

Components	Tables	e_t/e_0	μ/μ_N	m_g/m_e	S	N
Excited nucleon crystal $\downarrow nx^0_{G,1,0}$	10.1.1	0.00	-8.62546898	8.0000	0.0	1
Nucleon core $\uparrow nc^0_{1,1,0}$	10.2	0.00	+6.69957132	1918.0000	0.0	1
Basic lambda positron $\uparrow e^{+1}_{0,0,0}$	9.2	+1.00	+3.86797335	1.00	0.0	1
Oscillated lambda electron $\downarrow e^{-1}_{0,0,1}$	9.22	-1.00	-3.86797355	3.00	0.0	1
Excited oscillated neutron $\downarrow neo^0_{L1}$		**0.00**	**-1.92588766**	**1930.0000**	**0.0**	**1**
Measured values		0.00	-1.91304272	1838.68366	-	-
Calculated/measured ratio		1.00	1.0067	1.0497	-	-

Proton $\uparrow p^{+1}_{L1}$ of the ordinary spacetime level *L1* is made up of one basic golden nucleon crystal $\downarrow nx^0_{G,0,0}$, one nucleon core $\uparrow nc^0_{1,1,0}$ and one basic lambda positron $\uparrow e^{+1}_{0,0,0}$ as shown in Table 10.3.2 and by the equation:

$$\downarrow p^{+1}_{L1} = \downarrow nx^0_{G,0,0} + \uparrow nc^0_{1,1,0} + \uparrow e^{+1}_{0,0,0} \qquad (10.3\text{-}2)$$

Table 10.3.2. Components and properties of the proton $\uparrow p^{+1}_{L1}$ of the ordinary spacetime level *L1*.

Components	Tables	e_t/e_0	μ/μ_N	m_g/m_e	S	N
Basic nucleon crystal $\downarrow nx^0_{G,0,0}$	10.1.2	0.00	-7.73594709	8.00000	0.0	1
Nucleon core $\uparrow nc^0_{1,1,0}$	10.2	0.00	+6.69957132	1918.00000	0.0	1
Basic lambda positron $\uparrow e^{+1}_{0,0,0}$	9.2	+1.00	+3.86797355	1.000	0.0	1
Proton $\uparrow p^{+1}_{L1}$		**+1.00**	**+2.83159778**	**1927.00000**	**0.0**	**1**
Measured values		+1.00	+ 2.79284736	1836.15267	-	-
Calculated/measured ratio		1.00	1.0139	1.0495	-	-

Basic neutron$\downarrow nb_{L1}^{0}$ of the ordinary spacetime level *L1* is made up of basic golden nucleon crystal $\downarrow nx_{G,0,0}^{0}$, one nucleon core $\uparrow nc_{1,1,0}^{0}$, one combined basic ce-tron $\downarrow ce_{0,0,0}^{0}$ as shown in Table 10.3.3 and by the equation:

$$\downarrow nb_{L1}^{0} = \downarrow nx_{G,0,0}^{0} + \uparrow nc_{1,1,0}^{0} + \downarrow ce_{0,0,0}^{-1} \tag{10.3-3}$$

Table 10.3.3. Components and properties of the basic neutron $\downarrow nb_{L1}^{0}$ of the ordinary spacetime level *L1*.

Components	Tables	e_t/e_0	μ/μ_N	m_g/m_e	S	N
Basic nucleon crystal $\downarrow nx_{G,0,0}^{0}$	10.1.2	0.0	-8.62545898	8.0000	0.0	1
Nucleon core $\uparrow nc_{1,1,0}^{0}$	10.2	0.0	+6.69957132	1918.0000	0.0	1
Combined basic ce-tron $\downarrow ce_{0,0,0}^{0}$	9.17	0.00	-7.73594709	2.000000	0.0	1
Basic neutron $\downarrow nb_{L1}^{0}$		**0.0**	**-8.77232286**	**1928.0000**	**0.0**	**1**

Figure 10.3 shows the nucleon crystal structure in which the small circles indicate locations of the center 1 and the vertices 2 through 7 occupied by the nucleon constituent trons, while the empty circles indicate locations of the vertices 8 and 9 occupied by constituent trons of adjacent nucleons in complex nuclei.

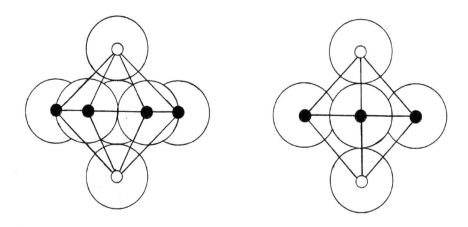

Figure 10.3. Front view (left) and side view (right) of a nucleon crystal structure.

10.4 Light Atoms & Isotopes of Ordinary Spacetime Level

Shown below are the structures and properties of light atoms and isotopes of the ordinary level *L1*.

Hydrogen atom $\downarrow_1^1 H_{L1}^0$ of the ordinary spacetime level *L1* is made up of one basic golden nucleon crystal $\downarrow nx_{G,0,0}^0$, one nucleon core $\uparrow nc_{1,1,0}^0$ and one combined excited ce-tron $\downarrow ce_{2,1,0}^0$ as shown in Table 10.4.1 and by the equation:

$$\downarrow_1^1 H_{L1}^0 = \downarrow nx_{G,0,0}^0 + \downarrow nc_{1,1,0}^0 + \downarrow ce_{2,1,0}^0 \qquad (10.4\text{-}1)$$

Table 10.4.1. Components and properties of hydrogen atom $\downarrow_1^1 H_{L1}^0$ of the ordinary spacetime level *L1*.

Components	Tables	e_t/e_0	μ/μ_N	m_g/m_e	S	N
Basic nucleon crystal $\downarrow nx_{G,0,0}^0$	10.1.2	0.00	-7.73594709	8.000000	0.0	1
Nucleon core $\downarrow nc_{1,1,0}^0$	10.2	0.00	+6.69957132	1918.000000	0.0	1
Combined excited ce-tron $\downarrow ce_{2,1,0}^0$	9.19	0.00	-1835.633640	2.000000	0.0	1
Hydrogen atom $\downarrow_1^1 H_{L1}^0$		**0.00**	**-1836.670016**	**1928.000000**	**0.0**	**1**
Measured values		0.00	-	1838.154061	-	-
Calculated/measured ratio		1.00	-	1.0489	-	-

Deuterium $\downarrow_1^2 H_{L1}^0$ of the ordinary spacetime level *L1* is made up of one hydrogen atom $\downarrow_1^1 H_{L1}^0$ and one basic neutron $\downarrow nb_{L1}^0$ as shown in Table 10.4.2, Fig. 10.4.1 and by the equation:

$$\downarrow_1^2 H_{L1}^0 = \downarrow_1^1 H_{L1}^0 + \downarrow nb_{L1}^0 \qquad (10.4\text{-}2)$$

Table 10.4.2. Components and properties of the deuterium $\downarrow_1^2 H_{L1}^0$ of the ordinary spacetime level *L1*.

Components	Tables	e_t/e_0	μ/μ_N	m_g/m_e	P_t	N
Hydrogen atom $\downarrow_1^2 H_{L1}^0$	10.4.1	0.00	-1836.670016	1928.000000	0.0	1
Basic neutron $\downarrow nb_{L1}^0$	10.3.3	0.00	-8.77232286	1928.000000	0.0	1
Deuterium $\downarrow_1^2 H_{L1}^0$		**0.00**	**-1845.442339**	**3856.000000**	**0.0**	**1**
Measured values		0.00	-	3671.485770	-	-
Calculated/measured ratio		1.00	-	1.0503	-	-

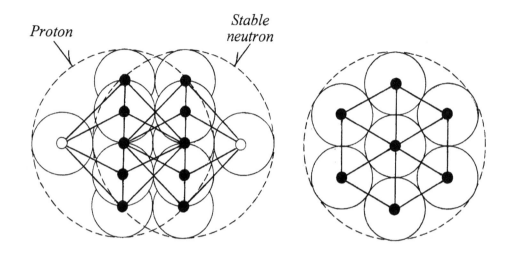

Figure 10.4.1. Structure of the deuterium nucleus:
front view (left) and side view (right).

Tritium $\downarrow_1^3 H_{L1}^0$ of the ordinary spacetime level $L1$ is made up of one hydrogen atom $\downarrow_1^1 H_{L1}^0$, one basic neutron $\downarrow nb_{L1}^0$ and one excited oscillated neutron $\downarrow neo_{L1}^0$ as shown in Table 10.4.3, Fig. 10.4.2 and by the equation:

$$\downarrow_1^3 H_{L1}^0 = \downarrow_1^1 H_{L1}^0 + \downarrow nb_{L1}^0 + \downarrow neo_{L1}^0 \tag{10.4-3}$$

Table 10.4.3. Components and properties of the tritium $\downarrow_1^3 H_{L1}^0$
of the ordinary spacetime level $L2$.

Components	Tables	e_t/e_0	μ/μ_N	m_g/m_e	S	N
Hydrogen atom $\downarrow_1^1 H_{L1}^0$	10.4.1	0.00	-1836.670016	1928.0000	0.0	1
Basic neutron $\downarrow nb_{L1}^0$	10.3.3	0.00	-8.77232286	1928.0000	0.0	1
Excited oscillated neutron $\downarrow neo_{L1}^0$	10.3.1	0.00	-1.92588766	1930.0000	0.0	**1**
Tritium $\downarrow_1^3 H_{L1}^0$		**0.00**	**-1847.368227**	**5786.0000**	**0.0**	**1**
Measured values		-	-	5497.9252	-	-
Calculated/measured ratio		-	-	1.0524	-	-

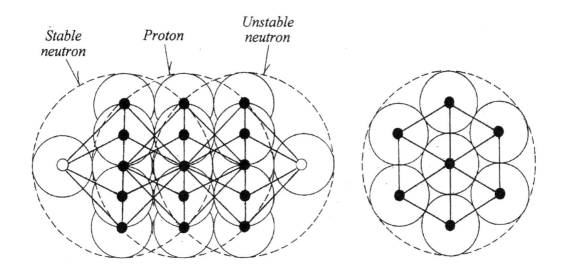

Figure 10.4.2. Structure of the tritium nucleus:
front view (left) and side view (right).

Helium-3 $\downarrow{}_2^3He_{L1}^0$ of the ordinary spacetime level *L1* is made up of two hydrogen atoms $\downarrow{}_1^1H_{L1}^0$ and $\uparrow{}_1^1H_{L1}^0$ with opposite signs of their magnetic moments and one basic neutron $\downarrow nb_{L1}^0$ as shown in Table 10.4.4, Fig. 10.4.3 and by the equation:

$$\downarrow{}_2^3He_{L1}^0 = \downarrow{}_1^1H_{L1}^0 + \uparrow{}_1^1H_{L1}^0 + \downarrow nb_{L1}^0 \tag{10.4-4}$$

Table 10.4.4. Components and properties of the helium-3 $\downarrow{}_2^3He_{L1}^0$
of the ordinary spacetime level *L1*.

Components	Table	e_t/e_0	μ/μ_N	m_g/m_e	P_t	N
Hydrogen atom $\downarrow{}_1^1H_{L1}^0$	10.4.1	0.00	-1836.670016	1928.00000	0.0	1
Hydrogen atom $\uparrow{}_1^1H_{L1}^0$	10.4.1	0.00	+1836.670016	1928.00000	0.0	1
Basic neutron $\downarrow nb_{L1}^0$	10.3.1	0.00	-8.77232286	1928.00000	0.0	1
Helium-3 $\downarrow{}_2^3He_{L1}^0$		**0.00**	**- 8.77232286**	**5784.00000**	**0.0**	**1**
Measured values		-	-	5497.88877	-	-
Calculated/measured ratio		-	-	1.0520	-	-

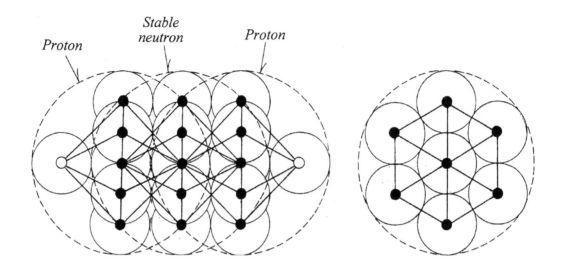

Figure 10.4.3. Structure of the helium-3 nucleus:
front view (left) and side view (right).

Helium-4 $^4_2He^0_{L1}$ of the spacetime level *L1* is made up of two hydrogen atoms $^1_1H^0_{L1}$ and two basic neutrons $\downarrow nb^0_{L1}$ with particles in both pairs having opposite signs of their magnetic moments as shown in Table 10.4.5, Fig. 10.4.4 and by the equation:

$$\downarrow^4_2He^0_{L1}=\downarrow^1_1H^0_{L1}+\uparrow H^0_{L1}+\downarrow nb^0_{L1}+\uparrow nb^0_{L1} \qquad (10.4\text{-}5)$$

Table 10.4.5. Components and properties of the helium-4 $^4_2He^0_{L1}$
of the ordinary spacetime level *L1*.

Components	Table	e_t/e_0	μ/μ_N	m_g/m_e	P_t	N
Hydrogen atom $\downarrow^1_1H^0_{L1}$	10.4.1	0.00	-1836.670016	1928.00000	0.0	1
Hydrogen atom $\uparrow^1_1H^0_{L1}$	10.4.1	0.00	+1836.670016	1928.00000	0.0	1
Stable neutron $\downarrow ns^0_{L1}$	10.3.1	0.00	-8.77232286	1928.00000	0.0	1
Stable neutron $\uparrow ns^0_{L1}$	10.3.1	0.00	+8.77232286	1928.00000	0.0	1
Helium-4 $^4_2He^0_{L1}$		**0.00**	**0.00**	**7712.00000**	**0.0**	**1**
Measured values	-	-	-	7296.30308	-	-
Calculated/measured ratio	-	-	-	1.0570	-	-

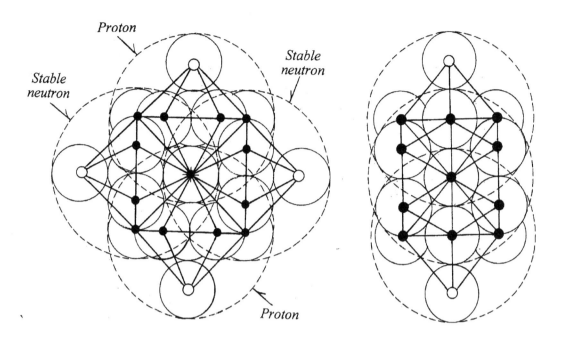

Figure 10.4.4. Structure of the helium-4 nucleus: front view (left) and side view (right).

10.5 Effects of Spacetime Levels of the Multiverse

The theory covers the Multiverse with six quantum spacetime levels ranging from *L0* to *L5* residing inside both the quantum vacuum infinity range (Fig. 10.5) and the quantum vacuum infinility range. The exponential excitation quantum states *m* of trons are dependent on the spacetime levels of the Multiverse and types of elementary particles as shown in Exhibit 10.5.1.

Exhibit 10.5.1. Quantum spacetime levels of the Multiverse.

Spacetime levels	Exponential excitation quantum states m of trons			
	Singulatron	**Ethertron**	**Electron**	**Positron**
L0	$m = 0$	$m = 0$	$m = 1$	$m = 1$
L1	**$m = 1$**	**$m = 1$**	**$m = 2$**	**$m = 2$**
L2	$m = 2$	$m = 2$	$m = 3$	$m = 3$
L3	$m = 3$	$m = 3$	$m = 4$	$m = 4$
L4	$m = 4$	$m = 4$	$m = 5$	$m = 5$
L5	$m = 5$	$m = 5$	$m = 5$	$m = 6$

Table 10.5.1 compares several physical properties of hydrogen atoms of the spacetime levels *L0, L1* and *L2* of the Multiverse.

Table 10.5.1. Relative parameters of the hydrogen atoms $\downarrow H_L^0$
at $n = 1$ of three spacetime levels L of the Multiverse.

Spacetime levels L of the Multiverse	Hydrogen atoms	Orbital electron radius ratio	Orbital electron magnetic moment	Hydrogen atom mass ratio	Density ratio	Modulus of elasticity
L0	$\downarrow H_{L0}^0$	1/137	1/11.76	1/80.33	2.58×10^8	3.53×10^8
L1	$\downarrow H_{L1}^0$	**1.0**	**1.0**	**1.0**	**1.0**	**1.0**
L2	$\downarrow H_{L2}^0$	137.00	11.71	136.30	3.89×10^{-7}	2.84×10^{-9}

As it follows from Table 10.5.1, as the spacetime level L of the Multiverse increases the orbital electron radius, the orbital electron magnetic moment and the mass of the hydrogen atom increase, while its density and modulus of elasticity decrease.

Table 10.5.2. Parameters of real negative toryces of hydrogen atoms in various spacetime levels L of the Multiverse.

Table 10.5.2. Parameters of real negative toryces of hydrogen atoms in various spacetime levels L of the Multiverse.

Toryx data	Spacetime levels of the Multiverse					
	L0	L1	L2	L3	L4	L5
m	1	2	3	4	5	6
n_{max}	9.6×10^8	2650	37	4	1	-
b_{max}	5.2717×10^{11}	5.2722×10^{11}	5.2099×10^{11}	3.6073×10^{11}	1.9305×10^{11}	-
b_u	5.275456×10^{11}					

It follows from Table 10.5.2, as the Multiverse level L increases the maximum numbers of the linear excitation quantum states n_{max} decrease to 1 at the spacetime level $L4$ of the Multiverse. Thus, the spacetime level $L4$ of the Multiverse is the highest spacetime level at which the hydrogen atoms are able to exist.

MAIN SUMMARY

- *Nucleons (protons and neutrons) are made up of three parts: nucleon crystal, nucleon core and nucleon leptons.*

- *Nucleon crystal is a bi-pyramid hexagonal crystal structure made up of intersecting golden trons retaining constituent trons of a nucleon core.*

- *Nucleon core is made up of excited lambda ethertrons and singulatrons.*

- *Nucleon leptons are made up of harmonic electrons and positrons.*

- *Isotopes and atoms are made up of nucleons and excited electrons.*

- *As the spacetime level L of the Multiverse increases the orbital electron radius, the orbital electron magnetic moment and the mass of the hydrogen atom increase, while its density.*

Notes:

11. SHORT-LIVED ELEMENTARY PARTICLES

11.1 Oscillated Leptons

According to the proposed theory, the oscillated leptons are short-lived oscillated electron and positron with masses significantly exceeding the mass of non-oscillated electron and positron. The oscillated leptons can be of two kinds: **excited lambda leptons** and **harmonic leptons**. Both these leptons exist in the oscillation quantum states described in Section 6.3.

Figure 11.1 shows a plot of a natural logarithm of the toryx oscillation factor $\ln Q_q$ as a function of the toryx oscillation quantum states q calculated from Eq. (6.3-1) in application to the excited lambda leptons of the ordinary matter $L1$.

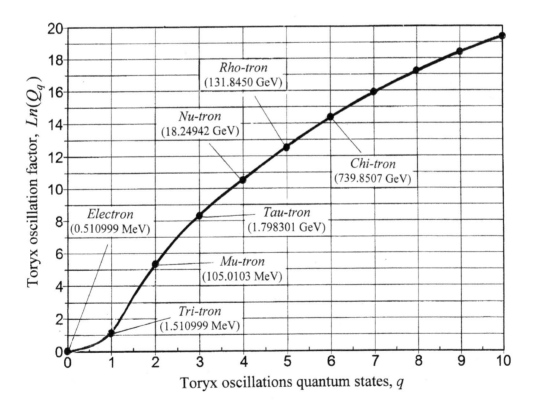

Figure 11.1. Toryx oscillation factor Q_q as a function of the toryx oscillation quantum states q.

Table 11.1 shows a comparison of calculated & measured properties of non-oscillated and oscillated excited lambda electrons and positrons (leptons) of the ordinary matter *L1*.

Table 11.1. Calculated & measured properties of non-oscillated and oscillated excited lambda electrons and positrons (leptons) of the ordinary matter *L1*

$$(b_1^+ = 0.50000666, \breve{b}_1^+ = 0.499999334, b_1^- = 37538.0, \breve{b}_1^- = -37537.0).$$

Leptons		Positive and negative leptons		μ_t / μ_B for leptons	
Name	Symbol	$m_{tg} / m_e = Q_q$	m_{tg}, GeV/c²	Positive	Negative
$e - tron$ $q = 0$	$e_{2,1,0}^{\pm 1}$	**1.000000**	**0.00051099893**	**+1.331643×10⁻⁵**	**-0.99973052**
Measured values:		1.000000	0.00051099893	-	-1.00115965
Calc./measured ratio:		1.0000	1.00	-	0.9986
$3e - tron$ $q = 1$	$e_{2,1,1}^{\pm 1}$	**3.000000**	**0.00151099893**	**+4.438808×10⁻⁶**	**-0.33324351**
Measured values:		-	-	-	-
Calc./measured ratio:		-	-	-	-
$\mu - tron$ $q = 2$	$e_{2,1,2}^{\pm 1}$	**205.500000**	**0.1050102797**	**+6.480012×10⁻⁸**	**-4.864869×10⁻³**
Measured values		-206.768284	0.1056583668	-	-4.841970×10⁻³
Calc./measured ratio		0.9939	0.9939	-	1.0047
$\tau - tron$ $q = 3$	$e_{2,1,3}^{\pm 1}$	**3519.18750**	**1.798301041**	**+3.783049×10⁻⁹**	**-2.840799×10⁻⁴**
Measured values		3478.18283	1.776990000	-	-
Calc./measured ratio		1.0120	1.0120	-	-
$v - tron$ $q = 4$	$e_{2,1,4}^{\pm 1}$	**35713.23612**	**18.2494254**	**+3.728709×10⁻¹⁰**	**-2.799328×10⁻⁵**
Measured values		-	-	-	-
Calc./measured ratio		-	-	-	-
$\rho - tron$ $q = 5$	$e_{2,1,5}^{\pm 1}$	**258014.1805**	**131.8449697**	**+5.161122×10⁻¹¹**	**-3.874712×10⁻⁶**
Measured values		244795.0427	125.0900000	-	-
Calc./measured ratio		1.0540 ?	1.0540 ?	-	-
$\chi - tron$ $q = 6$	$e_{2,1,6}^{\pm 1}$	**1447851.734**	**739.8506841**	**+9.197368×10⁻¹²**	**-6.904923×10⁻⁷**
Measured values		1468.713450	750 ?	-	-
Calc./measured ratio		0.98646758 ?	0.98646758 ?	-	-

It is follows from Table 11.1 and Figure 11.1:

- Calculated masses of *electron, mu-tron*, and *tau-tron* are very close to their respective measured values.
- Calculated magnetic moments of *electron* and *mu-tron* are very close to their respective measured values.
- Calculated mass of *rho-tron* is 5.4% greater than the mass of Higgs boson measured at CERN in 2017.

11.2 Basic Quarks

Quarks are the elementary particles having fractional charges. We proposed that they are made up of harmonic oscillated toryces. We consider below two groups of quarks:

Group 1 includes the quarks with the fractional charge $\varepsilon = +\frac{2}{3}e$. In the quarks of this group the relative radii of leading strings b_1 of one of their constituent toryces changes from 0.5 to $+\infty$.

Group 2 includes the quarks with the fractional charge $\varepsilon = -\frac{1}{3}e$. In the quarks of this group the relative radii of leading strings b_1 of one of their constituent toryces changes from -0.5 to $-\infty$.

Table 11.2. provides a comparison of calculated and measured relative masses of quarks. The empty cells indicate masses of predicted still undiscovered quarks.

Table 11.2. Comparison of calculated and measured relative masses of quarks.

q	Quark name	Quark group 1: charge $\varepsilon = +\frac{2}{3}e_0$ Ranges of relative masses m_g/m_e Calculated	Measured	Quark name	Quark group 2: charge $\varepsilon = -\frac{1}{3}e_0$ Ranges of relative masses m_g/m_e Calculated	Measured
0	-	0.34 - 0.85	-	-	0.34 - 1,36	-
1	Up	1.02 - 2.56	1.80 – 3.00	Down	1.02 - 4.09	4.5 - 5.3
2	-	70.0 - 175.0	-	Strange	70.0 – 280.0	90 – 100.0
3	Charm	1199 - 2997	1250 -1300	Bottom	1199 - 4798	4215 - 4690
4	-	12166 - 30416	-	-	12166 - 48665	-
5	Top	87897 - 219742	171099 - 174430	-	87897 - 351586	-

The relationship between the relative radii b_1' and b_1'' of leading strings of toryces making up the quarks is given by the equation:

$$b_1'' = \frac{b_1'}{b_1'(2+\varepsilon)-1} \tag{11.2-1}$$

where ε is the quark total relative charge.

11.3 Mesons

The mesons are comprised of three kinds of particles:

1. The neutral self-polarized harmonic electrons $e_{H,n,q}^0 = E_{H,n,q}^- + E_{H,n,q}^+$

2. The self-polarized harmonic electrons $ae_{H,n,q}^{-1} = A_{H,n,q}^- + E_{H,n,q}^-$

3. The self-polarized harmonic positrons $ae_{H,n,q}^{+1} = A_{H,n,q}^+ + E_{H,n,q}^+$.

Light unflavored mesons - Table 11.3.1 shows compositions and masses of the light unflavored mesons of the ordinary matter *L1*.

Table 11.3.1. Compositions and masses of the light unflavored mesons of the ordinary matter *L1*.

Constituent harmonic trons of mesons and their quantities					Meson relative mass m_g/m_e		
π^0	$e_{H,1,2}^0$	-	-	-	Calc.	Measured	Ratio
	1	-	-	-	274.00	264.14263	1.0373
$\pi^\pm$	$e_{H,1,2}^0$	$ae_{H,0,0}^{\pm1}$	-	-	Calc.	Measured	Ratio
	1	1	-	-	275.00	273.13204	1.0068
η^0	$e_{H,1,2}^0$	-	-	-	Calc.	Measured	Ratio
	4	-	-	-	1096.000	1072.1392	1.0223
$\rho^0(770)$	$e_{H,2,2}^0$	-	-	-	Calc.	Measured	Ratio
	5	-	-	-	1541.250	1517.596	1.0156

Strange mesons - Table 11.3.2 shows compositions and masses of the strange mesons of the ordinary matter *L1*.

Table 11.3.2. Compositions and masses of the strange mesons
of the ordinary matter *L1*.

	Constituent harmonic trons of mesons and their quantities				Meson relative mass m_g / m_e		
$K^{\pm}$	$e_{H,1,2}^{0}$	$ae_{H,1,2}^{\pm 1}$	-	-	Calc.	Measured	Ratio
	3	1	-	-	1027.000	966.102	1.064
K^{0}	$e_{H,1,2}^{0}$	$ae_{H,0,1}^{+1}$	$ae_{H,0,1}^{-1}$	-	Calc.	Measured	Ratio
	3	1	1	-	1030.500	973.806	1.058

Charmed mesons - Table 11.3.3 shows compositions and masses of the charmed mesons of the ordinary matter *L1*.

Table 11.3.3. Compositions and masses of the charmed mesons
of the ordinary matter L1.

	Constituent harmonic trons of mesons and their quantities				Meson relative mass m_g / m_e		
$D^{\pm}$	$e_{H,0,2}^{0}$	$e_{H,0,3}^{0}$	$ae_{H,0,1}^{\pm 1}$	-	Calc.	Measured	Ratio
	1	1	3	-	3727.688	3658.755	1.019
D^{0}	$e_{H,0,2}^{0}$	$e_{H,0,3}^{0}$	-	-	Calc.	Measured	Ratio
	1	1	-	-	3724.688	3649.401	1.021

Charm strange, mesons - Table 11.3.4 shows compositions and masses of the charmed, strange mesons of the ordinary matter *L1*.

Table 11.3.4. Compositions and masses of the charmed, strange mesons of the ordinary matter *L1*.

	Constituent harmonic trons of mesons and their quantities				Meson relative mass m_g / m_e		
$D_s^{\pm}$	$e_{H,0,2}^0$	$ae_{H,0,3}^{\pm 1}$	-	-	Calc.	Measured	Ratio
	2	1	-	-	3930.188	3852.239	1.020
$D_s^{*\pm}$	$e_{H,0,2}^0$	$ae_{H,0,3}^{\pm 1}$	-	-	Calc.	Measured	Ratio
	3	1	-	-	4135.688	4133.668	1.001
$D_{s0}^{*\pm}$	$e_{H,1,2}^0$	$ae_{H,0,3}^{\pm 1}$	-	-	Calc.	Measured	Ratio
	4	1	-	-	4615.188	4535.822	1.018
$D_{s1}^{\pm}$	$e_{H,3,2}^0$	$ae_{H,0,3}^{\pm 1}$	-	-	Calc.	Measured	Ratio
	4	1	-	-	4834.388	4813.317	1.004
$D_{s2}^{\pm}$	$e_{H,2,2}^0$	$ae_{H,0,3}^{\pm 1}$	-	-	Calc.	Measured	Ratio
	5	1	-	-	5060.438	5034.453	1.005

$C\bar{c}$ **mesons** - Table 11.3.5 shows compositions and masses of the $c\bar{c}$ mesons of the ordinary matter *L1*.

Table 11.3.5. Compositions and masses of the $c\bar{c}$ mesons of the ordinary matter *L1*.

	Constituent harmonic trons of mesons and their quantities				Meson relative mass m_g / m_e		
$\eta_c(1S)$	$e_{H,1,3}^0$	$ae_{H,0,2}^{+1}$	$ae_{H,0,2}^{-1}$	-	Calc.	Measured	Ratio
	1	3	3	-	5925.250	5832.302	1.016
$J/\psi(1S)$	$e_{H,2,3}^0$	$ae_{H,0,2}^{+1}$	$ae_{H,0,2}^{-1}$	-	Calc.	Measured	Ratio
	1	2	2	-	6100.781	6060.514	1.007
$\chi_{c0}(1P)$	$e_{H,1,3}^0$	$ae_{H,0,2}^{+1}$	$ae_{H,0,2}^{-1}$	-	Calc.	Measured	Ratio
	1	5	5	-	6747.250	6682.499	1.010
$\chi_{c1}(1P)$	$e_{H,2,3}^0$	$ae_{H,0,2}^{+1}$	$ae_{H,0,2}^{-1}$	-	Calc.	Measured	Ratio
	1	4	4	-	6922.781	6870.191	1.008

Bottom mesons - Table 11.3.6 shows compositions and masses of the bottom mesons of the ordinary matter *L1*.

Table 11.3.6. Compositions and masses of the bottom mesons
of the ordinary matter *L1*.

	Constituent harmonic trons of mesons and their quantities				Meson relative mass m_g / m_e		
$B^{\pm}$	$e^0_{H,0,3}$	$ae^{\pm 1}_{H,0,3}$	-	-	Calc.	Measured	Ratio
	2	1	-	-	10557.56	10331.04	1.022
B^0	$e^0_{H,0,3}$	$ae^{+1}_{H,0,3}$	$ae^{-1}_{H,0,0}$	-	Calc.	Measured	Ratio
	2	1	1	-	10557.56	10331.78	1.022

Bottom, charmed mesons - Table 11.3.7 shows compositions and masses of the bottom, charmed mesons of the ordinary matter *L1*.

Table 11.3.7. Compositions and masses of the bottom, charmed mesons
of the ordinary matter *L1*.

	Constituent harmonic trons of mesons and their quantities				Meson relative mass m_g / m_e		
B^+_c	$e^0_{H,1,3}$	$ae^{+1}_{H,0,3}$	-	-	Calc.	Measured	Ratio
	2	1	-	-	12903.688	12281.83	1.051
B^-_c	$e^0_{H,1,3}$	$ae^{-1}_{H,0,3}$	-	-	Calc.	Measured	Ratio
	2	1	-	-	12903.688	12281.83	1.051

$B\bar{b}$ **mesons** - Table 11.3.8 shows compositions and masses of the $b\bar{b}$ mesons of the ordinary matter *L1*.

Table 11.3.8. Compositions and masses of the $b\bar{b}$ mesons
of the ordinary matter *L1*.

Constituent harmonic trons of mesons and their quantities					Meson relative mass m_g/m_e		
Y(1S)	$e^0_{H,6,3}$	$ae^{+1}_{H,0,2}$	$ae^{-1}_{H,0,1}$	-	Calc.	Measured	Ratio
	3	1	1	-	18684.2344	18513.3461	1.009
Y(2S)	$e^0_{H,6,3}$	$ae^{+1}_{H,0,2}$	$ae^{-1}_{H,0,2}$	-	Calc.	Measured	Ratio
	3	3	3	-	19708.7344	19615.0314	1.005
Y(3S)	$e^0_{H,7,3}$	$ae^{+1}_{H,0,2}$	$ae^{-1}_{H,0,2}$	-	Calc.	Measured	Ratio
	3	4	4	-	20412.0000	20264.6218	1.007

11.4 Short-lived Baryons

Similarly to the protons and neutrons, each short-lived baryon is made up a nucleon crystal, nucleon core and leptons. Table 11.4.1 shows composition and properties of oscillated nucleon crystals of short-lived baryons. The number of ca-trons in each nucleon core of a short-lived baryon is greater than seven ca-trons making up the nucleon cores of protons and neutrons. Magnetic fields of the ca-trons are oriented in such a way that their total magnetic moments are reduced to zero ss shown in Table 11.4.2. Tables 11.4.3 – 11.4.9 show compositions and properties of several short-lived baryons of the ordinary matter *L1*.

Table 11.4.1. Composition and properties of the oscillated nucleon crystal $\downarrow nx^0_{0,0,1}$.

Leptons	Tables	Planes in Fig. 10.1.3	γ	e_t/e_0	μ/μ_N	m_g/m_e	S	N
Basic electron $e^{-1}_{0,0,0}$	9.1	A, B, C	90^0	-1.00	0.00	1.00	0.0	3
Basic positron $e^{+1}_{0,0,0}$	9.2	A, B, C	90^0	+1.00	0.00	1.00	0.0	3
Osc. electron $\downarrow e^{-1}_{0,0,1}$	9.22	D	0^0	-1.00	-3.86797355	3.00	0.0	1
Basic positron $\uparrow e^{+1}_{0,0,0}$	9.10	D	0^0	+1.00	+3.86797355	1.00	0.0	1
Oscillated nucleon crystal $\downarrow nx^0_{0,0,1}$			-	**0.00**	**0.00**	**10.00**	**0.0**	**1**

Table 11.4.2. The number N of the combined ca-trons $\updownarrow ca^0_{1,1,0}$ in nucleon cores and their total relative properties.

Short-lived baryons	N	m_g/m_e	μ/μ_N
$\Lambda^0, \Sigma^0, \Sigma^-$	8	2192.0	0.0
Σ^+, Ξ^-	9	2466.0	0.0
Ξ^0	10	2740.0	0.0
Ω^-	12	3288.0	0.0

Lambda-0 baryon Λ^0

Table 11.4.3. Components and properties of Lambda (0) baryon (Λ^0).

Components	Tables	e_t/e_0	μ/μ_N	m_g/m_e	S	N
Oscillated nucleon crystal $nx^0_{0,0,1}$	11.4.1	0.00	0.00	10.00000	0.0	1
Nucleon core $nc^0_{1,1,0}(\Lambda^0)$	11.4.2	0.00	0.00	2192.00000	0.0	1
Basic lambda positron $\uparrow e^+_{0,0,0}$	9.2	+1.00	+3.86797355	1.00000	0.0	1
Oscillated lambda electron $\downarrow e^-_{0,0,1}$	9.22	-1.00	-3.86797355	3.00000	0.0	1
Oscillated lambda ethertron $\downarrow a^0_{0,0,1}$	9.23	0.00	-0.64466226	1.500000	-2.0	1
Lambda-0 baryon Λ^0		**0.00**	**-0.64466226**	**2207.50000**	**-2.0**	**1**
Measured values		0.00	-0.61300000	2183.33726	-	1
Calculated/measured ratio		1.00	1.0517	1.0111	-	1

Sigma (+) baryon (Σ^+)

Table 11.4.4. Components and properties of Sigma (+) baryon (Σ^+).

Components	Tables	e_t/e_0	μ/μ_N	m_g/m_e	S	N
Oscillated nucleon crystal $nx_{0,0,1}^0$	11.4.1	0.00	0.00	10.00000	0.0	1
Nucleon core $nc_{1,1,0}^0(\Sigma^+)$	11.4.2	0.00	0.00	2466.00000	0.0	1
Basic lambda positron $\uparrow e_{0,0,0}^+$	9.2	+1.00	+3.86797355	1.00000	0.0	1
Oscillated lambda ethertron $\downarrow a_{0,0,1}^0$	9.23	0.00	-0.64466226	1.50000	-2.0	2
Sigma (+) baryon (Σ^+)		**0.00**	**+2.57864903**	**2480.0000**	**-4.0**	**1**
Measured values		0.00	+2.45864903	2327.53913	-	1
Calculated/measured ratio		1.00	1.0491	1.0655	-	1

Sigma (0) baryon (Σ^0)

Table 11.4.5. Components and properties of Sigma(0) baryon (Σ^0).

Components	Tables	e_t/e_0	μ/μ_N	m_g/m_e	S	N
Oscillated nucleon crystal $nx_{0,0,1}^0$	11.4.1	0.00	0.00	10.00000	0.0	1
Nucleon core $nc_{1,1,0}^0(\Sigma^0)$	11.4.2	0.00	0.00	2192.00000	0.0	1
Basic lambda positron $\uparrow e_{0,0,0}^+$	9.2	+1.00	+3.86797355	1.00000	0.0	1
Oscillated lambda electron $\downarrow e_{0,0,2}^-$	-	-1.00	-0.05646677	205.50000	0.0	1
Basic lambda ethertron $\downarrow a_{0,0,0}^0$	9.3	0.00	-1.93398677	0.5000000	$-\frac{2}{3}$	1
Sigma (0) baryon (Σ^0)		**0.00**	**-1.87752001**	**2409.00000**	$\mathbf{-\frac{2}{3}}$	**1**
Measured values		0.00	-1.61000000	2333.94227	-	1
Calculated/measured ratio		1.00	1.1662	1.0322	-	1

Sigma (-) baryon (Σ^-)

Table 11.4.6. Components and properties of Sigma(-) baryon (Σ^-).

Components	Tables	e_t/e_0	μ/μ_N	m_g/m_e	S	N
Oscillated nucleon crystal $nx_{0,0,1}^0$	11.4.1	0.00	0.00	10.00000	0.0	1
Nucleon core $nc_{1,1,0}^0(\Sigma^-)$	11.4.2	0.00	0.00	2192.00000	0.0	1
Oscillated lambda electron $\downarrow e_{0,0,2}^{-1}$	-	-1.00	-0.05646677	205.50000	0.0	1
Oscillated lambda ethertron $\downarrow a_{0,0,1}^0$	9.23	0.00	-0.64466226	1.50000	-2.0	2
Sigma (-) baryon (Σ^-)		**0.00**	**-1.345791**	**2410.50000**	**-4.0**	**1**
Measured values		0.00	-1.160000	2343.34934	-	1
Calculated/measured ratio		1.00	1.1602	1.0287	-	1

Ksi (0) baryon (Ξ^0)

Table 11.4.7. Components and properties of Ksi (0) baryon (Ξ^0).

Components	Tables	e_t/e_0	μ/μ_N	m_g/m_e	S	N
Oscillated nucleon crystal $nx_{0,0,1}^0$	11.4.1	0.00	0.00	10.00000	0.0	1
Nucleon core $nc_{1,1,0}^0(\Xi^0)$	11.4.2	0.00	0.00	2740.00000	0.0	1
Oscillated lambda ethertron $\downarrow a_{0,0,1}^0$	9.23	0.00	-0.64466226	1.500000	-2.0	2
Ksi (0) baryon (Ξ^0)		**0.00**	**-1.289325**	**2753.00000**	**-4.0**	**1**
Measured values		0.00	-1.250000	2573.11694	-	1
Calculated/measured ratio		1.00	1.0315	1.0699	-	1

Ksi (-) baryon (Ξ^-)

Table 11.4.8. Components and properties of Ksi (-) baryon (Ξ^-).

Components	Tables	e_t/e_0	μ/μ_N	m_g/m_e	S	N
Oscillated nucleon crystal $nx^0_{0,0,1}$	11.4.1	0.00	0.00	10.00000	0.0	1
Nucleon core $nc^-_{1,1,0}(\Xi^-)$	11.4.2	0.00	0.00	2466.00000	0.0	1
Oscillated lambda electron $\downarrow e^{-1}_{0,0,2}$	-	-1.00	-0.05646677	205.50000	0.0	1
Oscillated lambda ethertron $\downarrow a^0_{0,0,1}$	9.23	0.00	-0.64466226	1.50000	-2.0	1
Ksi (-) baryon (Ξ^-)		**0.00**	**-0.70112902**	**2683.00000**	**-2.0**	**1**
Measured values		0.00	-65070000	2586.52206	-	1
Calculated/measured ratio		1.00	1.0775	1.0373	-	1

Omega (-) baryon (Ω^-)

Table 11.4.9. Components and properties of Omega (Ω^-).

Components	Tables	e_t/e_0	μ/μ_N	m_g/m_e	S	N
Oscillated nucleon crystal $nx^0_{0,0,1}$	11.4.1	0.00	0.00	10.000000	0.0	1
Nucleon core $nc^0_{1,1,0}(\Sigma^0)$	11.4.2	0.00	0.00	3288.000000	0.0	1
Oscillated lambda electron $\downarrow e^-_{0,0,2}$	-	-1.00	-0.05646677	205.500000	0.0	1
Basic lambda ethertron $\downarrow a^0_{0,0,1}$	9.3	0.00	-1.93398677	0.500000	$-\frac{2}{3}$	1
Omega (-) baryon (Ω^-)		**-1.00**	**-1.99045354**	**3504.000000**	$-\frac{2}{3}$	**1**
Measured values		-1.00	-2.02000000	3272.90315	-	1
Calculated/measured ratio		1.00	0.9854	1.0706	-	1

MAIN SUMMARY

This chapter shows calculated properties of both discovered and predicted short-lived elementary particles, including:

> *7 oscillated leptons (see Table 11.1),*
> *12 basic quarks (see Table 11.2),*
> *25 mesons (see Tables 11.3.1 – 11.3.8) and*
> *5 short-lived baryons (see Tables 11.4.1 – 11.4.9).*

It also provides a comparison between calculated and measured particle masses.

<u>*Notes*</u>

12. CELESTIAL BODIES & FIELDS OF THE MACRO-WORLD

CONTENTS

12.1 Basic Structure & Parameters of Macro-Toryces

Celestial bodies are formed after myriads of atoms of the micro-world are joined together. Associated with each celestial body are the *macro-toryces* forming toroidal celestial spacetime fields responsible for interactions between celestial bodies. Figure 12.1 shows a macro-toryx associated with the central body A and encompassing the satellite body B.

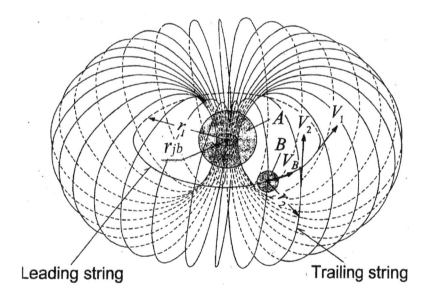

Leading string   Trailing string

Figure 12.1. Central body A with a macro-toryx encompassing the satellite body B.

Similar to the micro-toryx (Figs. 1.1.1 – 1.1.3), the macro-toryx contains two strings, a *leading string* and a *trailing string*. Leading string is double-circular; it is moving with the velocity V_1 along a circle with the radius r_1. Trailing string is double-toroidal with the radius r_2 in which each branch propagates along its toroidal spiral path with the spiral velocity V_2 that has two components, the translational velocity V_{2t} and the rotational velocity V_{2r}.

12.2 Macro-Toryx Genetic Codes

Exhibit 12.2.1 shows the macro-toryx genetic codes expressed in absolute spacetime units.   These genetic codes are the same as those used for micro-toryces shown in Exhibit 1.3.

Exhibit 12.2.1. Macro-toryx genetic code in absolute units.

- The length of one winding of trailing string L_2 is equal to the length of one winding of leading string L_1:

$$L_2 = L_1 = 2\pi r_1 \qquad (12.2\text{-}1)$$

- The macro-toryx eye radius r_{jb} is constant:

$$r_{jb} = r_1 - r_2 = const. \qquad (12.2\text{-}2)$$

- The spiral velocity of trailing string V_2 is constant at each point of its spiral path:

$$V_2 = \sqrt{V_{2t}^2 + V_{2r}^2} = c = const. \qquad (12.2\text{-}3)$$

Similarly to the micro-toryx, the macro-toryx genetic codes can also be simplified (Exhibit 12.2.2) by expressing the macro-toryx spacetime parameters in relative units in respect to the constant macro-toryx parameters: the eye radius r_{jb} and the velocity of light c.

Exhibit 12.2.2. Macro-toryx genetic codes in relative units.

- The relative length of one winding of trailing string l_2 is equal to the relative length of one winding of leading string l_1:

$$l_2 = l_1 \qquad (12.2\text{-}4)$$

- The macro-toryx relative eye radius b_{jb} is equal to 1:

$$b_{jb} = b_1 - b_2 = 1 \qquad (12.2\text{-}5)$$

- The relative spiral velocity of trailing string β_2 is equal to 1 at each point of its spiral path:

$$\beta_2 = \sqrt{\beta_{2t}^2 + \beta_{2r}^2} = 1 \qquad (12.2\text{-}6)$$

12.3 Classification of Macro-Toryces & Macro-Trons

Similarly to the classification of the micro-toryces shown in Chapter 3, the macro-toryces are also divided into four groups as shown in Fig. 12.3:

- Real negative macro-toryces
- Real positive macro-toryces
- Imaginary positive macro-toryces
- Imaginary negative macro-toryces.

Similarly to the classification of the micro-trons shown in Section 8.2, the macro-trons are also divided into four groups: macro-electrons, macro-positrons, macro-ethertrons and macro-singulatrons.

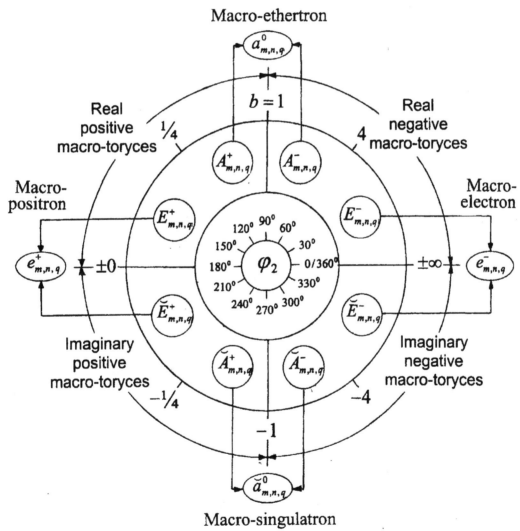

Figure 12.3. Formation of four macro-trons from polarized macro-toryces.

The structures of the macro-toryces and macro-trons are similar to the structures of the micro-toryces and micro-trons shown Section 9.

12.4 Macro-Toryx Law of Planetary Motion

We described In Section 1.7 the law of planetary system applicable to the micro-toryces by the equation:

$$\beta_1 = \frac{V_1}{c} = \frac{\sqrt{2b_1 - 1}}{b_1} \tag{12.4-1}$$

The above two equations also describe the **macro-toryx law of planetary motion** with only one difference. In the planetary system applicable to the micro-toryces b_1 is expressed by Eq. (1.6-1a), whereas in the macro-toryx law of planetary motion b_1 is expressed by the equation:

$$b_1 = \frac{r_1}{r_{jb}} \tag{12.4-2}$$

where r_{0m} is the eye radius of the macro-toryx associated with the body A with the mass m_A (Fig. 12.1) that is equal to:

$$r_{jb} = \frac{m_A G k_b}{2c^2} \tag{12.4-3}$$

where k_b is the constant celestial body parameter.

For the case when $b_1 \gg 1$, the above equation reduces to the form:

$$\beta_1 = \frac{V_1}{c} = \sqrt{\frac{2}{b_1}} = \sqrt{\frac{2r_{jb}}{r_1}} \tag{12.4-4}$$

where G is the Newtonian constant of gravitation.

From Eqs. (12.4-1), (12.4-3) and (12.4-4), it is possible to express the macro-toryx law of planetary motion in the form:

$$r_1^3 = k T_1^2 \left(1 - \frac{r_{jb}}{2r_1} \right) \tag{12.4-5}$$

Consequently, for the case when r_1 is much greater than the macro-toryx eye radius r_{jb}, Eq. (12.4-5) reduces to the Kepler's third law of planetary motion:

$$r_1^3 = k T_1^2 \tag{12.4-6}$$

12.5 Interactions Between Celestial Bodies

The macro-trons associated with celestial bodies are responsible for interactions between celestial bodies. Figure 12.5 shows two adjacent celestial bodies A and B and the respective macro-toryces A and B associated with these bodies. The macro-toryx A represents both real and imaginary negative macro-toryces of the macro-electron associated with the body A, while the macro-toryx B represents both real and imaginary negative macro-toryces of the macro-electron associated with the body B. Similarly, both real and imaginary macro-toryces of macro-positrons are also associated with the bodies A and B. They all are located inside their respective toryx eyes. Consequently, the leading strings of the macro-toryces of the macro-electron associated with the body A intersect with the macro-positron associated with the body B, while the leading strings of the macro-toryces of the macro-electron associated with the body B intersect with the macro-positron associated with the body A.

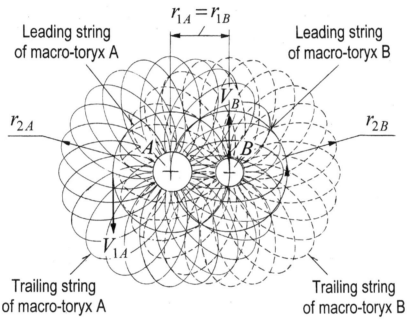

Figure 12.5. Two adjacent celestial bodies A and B with macro-toryces A and B associated with these bodies.

Let the macro-toryx A to be associated with the body A having the mass m_A. The macro-toryx circular leading string with the radius r_{1A} propagates around a center of the body A with the orbital velocity V_{1A}, while its toroidal trailing string with the radius r_{2A} propagates synchronously with the leading string. Located at the distance $r_{1B} = r_{1A}$ from the center of the body A is a center of the body B having the mass m_B. The body B moves around the center of the body A along a circular path with the radius r_{1B} with the orbital velocity V_B.

The behavior of the body B depends on a relationship between its orbital velocity V_B and the velocity of leading string V_{1A} of the macro-toryx A associated with the body A. When these velocities are not exactly the same, it   means the body B does not follow the macro-toryx law of planetary motion described by Eq. (12.4-1). Consequently, the body B will move in a radial direction either towards to or away from the body A with the *spacetime acceleration* a_{bs} described by the proposed equation shown in Exhibit 12.5.

Exhibit 12.5. Spacetime acceleration of a body.

Spacetime acceleration a_{bs} of the body B in respect to a center of the body A is equal to:

$$a_{bs} = \frac{V_{1A}^2 - V_B^2}{r_{1A}} = \frac{V_{1A}^2(1 - \gamma_V^2)}{r_{1A}} \tag{12.5-1}$$

where γ_V is the velocity ratio that is equal to:

$$\gamma_V = V_B / V_1 \tag{12.5-2}$$

Considering Eqs. (12.4-1), (12.4-4) and (12.5-1), the spacetime acceleration a_{bs} of the body B in respect to the body A separated by the distance b_{1A} is equal to:

$$a_{bs} = \frac{m_A G k_b}{r_{1A}^2} \frac{2b_{1A} - 1}{2b_{1A}}(1 - \gamma_V^2) \tag{12.5-3}$$

For a particular case when $b_{1A} \gg 1$ and $\gamma_V = 0$, Eq. (12.5-3) reduces to the form:

$$a_b = \frac{m_A G k_b}{r_{1A}^2} \tag{12.5-4}$$

Consequently, when the orbital velocity V_B of the body B around a center of the body A is the same as the velocity of propagation of leading string V_{1A} of the macro-toryx associated with the body A ($\gamma_V = 1$), the body B will orbit the body A according to the spacetime law of planetary motion and, consequently, the velocity V_B and its distance to the body B will remain unchanged. When the velocity V_B of the body B around the body A is different than the velocity of propagation of leading string V_1 of the macro-toryx associated with the body A, the body B accelerates either towards to or away from the body A.

12.6   Quantum States of Macro-Toryces of Celestial Bodies

Tables 12.6.1 – 12.6.5 show parameters of planets and moons in our solar system, and also their possible past, current and projected future locations.

Table 12.6.1. Parameters and status of planets in our solar system.

n	Planets	Measured distance to Sun r_{1m}, m	Relative measured dist. to Sun b_{1m}	Relative calculated dist. to Sun b_{1c}	Ratio b_{1m}/b_{1c}	Status of Sun's planets
1	Sun1	6.2378×10^{09}	8.4458×10^{6}	8.4458×10^{6}	1.0000	Former planets swallowed by the Sun
2	Sun 2	2.4951×10^{10}	3.3784×10^{7}	3.3784×10^{7}	1.0000	
3	**Mercury**	5.7909×10^{10}	7.8408×10^{7}	7.6014×10^{7}	**1.0315**	**Currently existing**
4	**Venus**	1.0821×10^{11}	1.4651×10^{8}	1.3514×10^{8}	**1.0842**	
5	**Earth**	1.4960×10^{11}	2.0255×10^{8}	2.1115×10^{8}	**0.9593**	
6	**Mars**	2.2794×10^{11}	3.0863×10^{8}	3.0406×10^{8}	**1.0150**	
7	Hungaria	3.0567×10^{11}	4.1387×10^{8}	4.1387×10^{8}	1.0000	Future planets that could be formed from asteroid belts
8	Phocaea	3.9923×10^{11}	5.4055×10^{8}	5.4055×10^{8}	1.0000	
9	Cybele	5.0527×10^{11}	6.8412×10^{8}	6.8412×10^{8}	1.0000	
10	Hilda	6.2382×10^{11}	8.4464×10^{8}	8.4464×10^{8}	1.0000	
11	**Jupiter**	7.7833×10^{11}	1.0538×10^{9}	1.0220×10^{9}	**1.0312**	**Currently existing**
12	Io	8.9826×10^{11}	1.2162×10^{9}	1.2162×10^{9}	1.0000	Former planets captured by the Jupiter and became its moons
13	Europa	1.0542×10^{12}	1.4273×10^{9}	1.4273×10^{9}	1.0000	
14	Ganymede	1.2227×10^{12}	1.6555×10^{9}	1.6555×10^{9}	1.0000	
15	**Saturn**	1.4270×10^{12}	1.9321×10^{9}	1.9004×10^{9}	**1.0167**	**Currently existing**
16	Tethys	1.5970×10^{12}	1.1622×10^{9}	1.1622×10^{9}	1.0000	Former planets captured by the Saturn and became its moons
17	Dione	1.1029×10^{12}	2.4410×10^{9}	2.4410×10^{9}	1.0000	
18	Rhea	2.0210×10^{12}	2.7364×10^{9}	2.7364×10^{9}	1.0000	
19	Titan	2.2518×10^{12}	3.0489×10^{9}	3.0489×10^{9}	1.0000	
20	Iapetus	2.4951×10^{12}	3.3783×10^{9}	3.3783×10^{9}	1.0000	
21	**Uranus**	2.8696×10^{12}	3.8853×10^{9}	3.7247×10^{9}	**1.0431**	**Currently existing**
22	Miranda	3.0190×10^{12}	4.0877×10^{9}	4.0877×10^{9}	1.0000	Former planets captured by the Uranus and became its moons
23	Ariel	3.2998×10^{12}	4.4678×10^{9}	4.4678×10^{9}	1.0000	
24	Umbriel	3.5929×10^{12}	4.8647×10^{9}	4.8647×10^{9}	1.0000	
25	Titania	3.8988×10^{12}	5.2789×10^{9}	5.2789×10^{9}	1.0000	
26	Oberon	4.2170×10^{12}	5.7097×10^{9}	5.7097×10^{9}	1.0000	
27	**Neptune**	4.4966×10^{12}	6.0883×10^{9}	6.1572×10^{9}	**0.9888**	**Currently existing**
28	Proteus	4.8904×10^{12}	6.6215×10^{9}	6.6214×10^{9}	1.0000	Former planets captured by the Neptune and became its moons
29	Triton	5.2460×10^{12}	7.1029×10^{9}	7.1029×10^{9}	1.0000	

Table 12.6.2. Parameters and status of the Jupiter's moons.

n	Jupiter's moons	Measured distance to Jupiter r_{1m}, m	Relative measured distance to Jupiter b_{1m}	Relative calculated distance to Jupiter b_{1c}	Ratio b_{1m}/b_{1c}	Status of Jupiter's moons
1	Jupiter 1	2.7115×10^7	3.8439×10^7	3.8439×10^7	1.0000	Former moons swallowed by the Jupiter
2	Jupiter 2	1.0846×10^8	1.5376×10^8	1.5376×10^8	1.0000	
3	Io	2.6200×10^8	3.7143×10^8	3.4595×10^8	1.0737	Currently existing moons
4	Europa	4.1690×10^8	5.9103×10^8	6.1502×10^8	0.9610	
5	Ganymede	6.6490×10^8	9.4261×10^8	9.6097×10^8	0.9809	
7	Callisto	1.1701×10^9	1.6588×10^9	1.8835×10^9	0.8807	

Table 12.6.3. Parameters and status of the Saturn's moons.

n	Saturn's moons	Measured distance to Saturn r_{1m}, m	Relative measured distance to Saturn b_{1m}	Relative calculated distance to Saturn b_{1c}	Ratio b_{1m}/b_{1c}	Status of Saturn's moons
1	Saturn 1	2.3974×10^7	1.1355×10^8	1.1355×10^8	1.0000	Former moons swallowed by the Saturn
2	Saturn 2	9.5890×10^7	4.5421×10^8	4.5421×10^8	1.0000	
3	Tethys	2.9462×10^8	1.3955×10^9	1.0220×10^9	1.3655	Currently existing moons
4	Dione	3.7740×10^8	1.7876×10^9	1.8168×10^9	0.9839	
5	Rhea	5.2711×10^8	2.4967×10^9	2.8388×10^9	0.8795	
7	Titan	1.2219×10^9	5.7877×10^9	5.5641×10^9	1.0402	
12	Iapetus	3.5608×10^9	1.6866×10^{10}	1.6352×10^{10}	1.0315	

Table 12.6.4. Parameters and status of the Uranus' moons.

N	Uranus' moons	Measured distance to Uranus r_{1m}, m	Relative measured distance to Uranus b_{1m}	Relative calculated distance to Uranus b_{1c}	Ratio b_{1m}/b_{1c}	Status of Uranus' moons
1	Uranus 1	1.2111×10^7	3.7538×10^8	3.7538×10^8	1.0000	Former moons swallowed by the Uranus
2	Uranus 2	4.8446×10^7	1.5015×10^9	1.5015×10^9	1.0000	
3	Miranda	1.2978×10^8	4.0225×10^9	3.3784×10^9	1.1907	Currently existing moons
4	Ariel	1.9102×10^8	5.9207×10^9	6.0061×10^9	0.9858	
5	Umbriel	2.6630×10^8	8.2540×10^9	9.3845×10^9	0.8795	
6	Titania	4.3591×10^8	1.3511×10^{10}	1.3514×10^{10}	0.9998	
7	Oberon	5.8352×10^8	1.8086×10^{10}	1.8394×10^{10}	0.9833	

Table 12.6.5. Parameters and status of the Neptun's moons.

n	Neptune's moons	Measured distance to Neptune r_{1m}, m	Relative measured distance to Neptune b_{1m}	Relative calculated distance to Neptune b_{1c}	Ratio b_{1m}/b_{1c}	Status of Neptune's moons
1	Neptune 1	1.4352×10^7	3.7538×10^8	3.7538×10^8	1.0000	Former moons swallowed by the Neptune
2	Neptune 2	4.8446×10^7	1.5015×10^9	1.5015×10^9	1.0000	
3	Proteus	1.1765×10^8	3.0772×10^9	3.3784×10^9	0.9108	**Currently existing moons**
5	Triton	3.5476×10^8	9.2791×10^9	9.3845×10^9	0.9888	

Exhibit 6.1 describes the quantization parameter z used for the micro-toryces. Similarly, the quantum states of macro-toryces of celestial bodies are described by the quantization parameter Z shown in Exhibit 12.6.

Exhibit 12.6. The quantization parameter Z of a macro-toryx.

The quantization parameter Z of a macro-toryx is expressed by the equation:

$$b_1 = Z = 2(n\Lambda)^2 \qquad\qquad (12.6\text{-}1)$$

Table 12.6.6 shows the values of the celestial body parameters k_b, the body masses m_A, the equatorial body radii r_b and the body eye radii r_{jb} for the largest bodies of our solar system.

Table 12.6.6. Parameters of largest celestial bodies of our solar system.

Celestial body	k_b	m_{A}, kg	r_b, m	r_{jb}, m
Sun	225	1.98910×10^{30}	6.9550×10^8	166177.48
Jupiter	1024	1.89973×10^{27}	7.1492×10^7	722.31219
Saturn	3025	5.68598×10^{26}	6.0268×10^7	638.65010
Uranus	10000	8.68910×10^{25}	2.5559×10^7	322.63000
Neptune	10000	1.02966×10^{26}	2.4764×10^7	382.32000

Table 12.6.7 shows parameters and status of objects most distant from the Sun.

Table 12.6.7. Parameters and status of objects most distant from the Sun.

n	Objects	Measured distance to Sun r_{1b}, AU	Measured distance to Sun r_{1b}, m	Relative measured dist. to Sun b_{1m}	Relative calculated dist. to Sun b_{1c}	Ratio b_{1m}/b_{1c}
30	Predicted 1	37.5279	5.6142×10^{12}	3.3784×10^7	3.3784×10^7	1.0000
31	Pluto	39.5000	5.9092×10^{12}	3.5560×10^7	3.6074×10^7	0.9857
32	Haumea	43.1283	6.4520×10^{12}	3.8826×10^7	3.8439×10^7	1.0101
33	Makemake	45.7888	6.8500×10^{12}	4.1221×10^7	4.0879×10^7	1.0064
34	Predicted 2	48.2025	7.2111×10^{12}	4.3394×10^7	4.2294×10^7	1.0000
35	2005 QU182	51.3230	7.6779×10^{12}	4.6203×10^7	4.5984×10^7	1.0048
36	2004 VU130	53.8000	8.0485×10^{12}	4.8433×10^7	4.8649×10^7	0.9956
37	2013 JQ64	58.4000	8.7366×10^{12}	5.2574×10^7	5.1390×10^7	1.0231
38	2003 QX113	59.9000	8.9410×10^{12}	5.2574×10^7	1.2196×10^{10}	0.9948
39	Predicted 3	63.4221	9.4880×10^{12}	5.7095×10^7	5.7095×10^7	1.0000
40	Eris	67.6471	1.0120×10^{13}	6.0899×10^7	6.0061×10^7	1.0140
41	2010 GB174	71.9000	1.0756×10^{13}	6.4727×10^7	6.3171×10^7	1.0258
42	Predicted 4	73.5547	1.1004×10^{13}	6.6217×10^7	6.6217×10^7	1.0000
43	2015 TJ367	78.7000	1.1774×10^{13}	7.0849×10^7	6.9408×10^7	1.0208
44	2015 UH87	82.7000	1.2372×10^{13}	7.2109×10^7	7.2674×10^7	1.0244
45	2014 FC69	84.7000	1.2671×10^{13}	7.6271×10^7	7.6014×10^7	1.0031
46	2015 TH387	89.1000	1.3329×10^{13}	8.0212×10^7	7.9430×10^7	1.0098
47	2014 UZ224	90.9000	1.3599×10^{13}	8.1832×10^7	8.2921×10^7	0.9869
48	Eris	96.1000	1.4377×10^{13}	8.6513×10^7	8.6488×10^7	1.0003
49	Predicted 5	100.1161	1.4977×10^{13}	9.0129×10^7	9.0129×10^7	1.0000
50	Predicted 6	104.2442	1.5595×10^{13}	9.3845×10^7	9.3845×10^7	1.0000
51	Predicted 7	108.4556	1.622×10^{13}	9.7636×10^7	9.7636×10^7	1.0000
52	Predicted 8	112.7505	1.6867×10^{13}	1.0150×10^8	1.0150×10^8	1.0000
53	Predicted 9	117.1287	1.7522×10^{13}	1.0544×10^8	1.0544×10^8	1.0000
54	Predicted 10	121.5904	1.8190×10^{13}	1.0946×10^8	1.0946×10^8	1.0000
55	2018 VG18	125.0000	1.8700×10^{13}	1.1253×10^8	1.1253×10^8	0.9910
56	Predicted 11	130.7639	1.9562×10^{13}	1.1772×10^8	1.1772×10^8	1.0000
57	Predicted 12	135.4757	2.0267×10^{13}	1.2196×10^8	1.2196×10^8	1.0000
58	FarFarOut	140.0000	2.0944×10^{13}	1.2603×10^8	1.2628×10^8	0.9981

12.7 Galaxy Law of Star Motion

Consider a star moving along a circular path with the radius r_s around a galaxy center with the orbital velocity V_s as shown in Fig. 12.7.

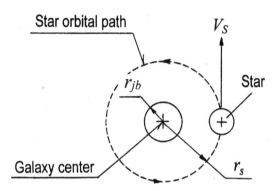

Figure 12.7. A star moving around a galaxy center.

According to the macro-toryx law of planetary motion described by Eqs. (12.4-1) – (12.4-4) the orbital velocities of all planets are dependent on the constant macro-toryx eye radius r_{jb} of the central body, such as the Sun of our solar system. Consequently, the velocity of each planet was inversely proportional to the square root of the planet orbital radius r_1.

Many astronomers expected that a similar relationship must exist between orbital velocities and radii of the stars in the galaxies, but numerous astronomical observations established that the star orbital velocities V_s are not dependent on the star orbital radius r_s. To explain this discrepancy, the astronomers proposed that the galaxies contain a so-called ***dark matter***.

Exhibit 12.7. The galaxy law of Star Motion.

- The eye radius r_{jg} of a macro-toryx associated with a star is directly-proportional to the galaxy mass m_G contained inside a sphere with the radius equal to the star orbital radius r_s:

$$r_{jg} = \frac{k_g m_G G}{2c^2} \qquad (12.7\text{-}1)$$

- The ratio s of the galaxy mass m_G contained inside a sphere with the radius equal to the star orbital radius r_s is constant:

$$\frac{m_G}{r_s} = s = const. \qquad (12.7\text{-}2)$$

Where k_g = coefficient related to a specific galaxy.

The Galaxy Law of Star Motion provides an alternative explanation why the star orbital velocities V_s are independent on the star orbital radii r_s. This is accomplished without any references to the dark matter. It follows from this law presented in Exhibit 12.7 that the star relative orbital radius is constant:

$$b_s = \frac{r_s}{r_{jg}} = \frac{2c^2}{sGk_g} = const.$$

(12.7-3)

Consequently, the star relative orbital velocity is also constant:

$$\beta_s = \frac{V_s}{c} = \frac{\sqrt{2b_s - 1}}{b_s} = const.$$

(12.7-4)

For the case when b_s is much greater than 1, Eq. (12.7-4) reduces to the form:

$$\beta_s = \frac{V_s}{c} = \sqrt{\frac{2}{b_s}} = \sqrt{sG} = const.$$

(12.4-5)

12.8 Classification of Stars

Depending on the relationship between the outer radius r_b of a star and the eye radius r_{jb} of its macro-toryx, the stars can be divided into three kinds: ***outverted, inverted*** and ***imaginary*** as shown in Figure 12.8.

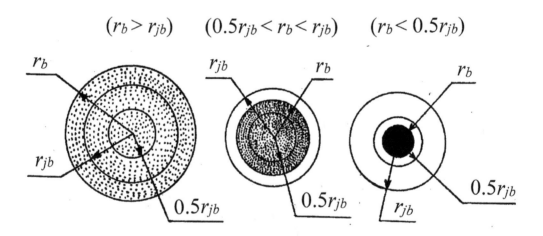

Figure 12.8. Three kinds of stars: outverted (left), inverted (center) and imaginary (right).

Outverted stars $(r_b \gg r_{jb})$ – In the outverted stars, the outer radius r_b of the star body is much greater than the eye radius r_{jb} of a macro-toryx associated with this star. ***Our Sun*** is an example of outverted stars.

Marginally-outverted stars $(r_b > r_{jb})$ – In the marginally-outverted stars, the outer radius r_b of the star body is slightly greater than the eye radius r_{jb} of a macro-toryx associated with this star. *Neutron stars* and *pulsars* are examples of marginally-outverted stars.

Inverted stars $(0.5r_{jb} < r_b < r_{jb})$ – In the inverted stars, the outer radius r_b of a star body is greater than one half of the eye radius r_{jb}, but less than the eye radius r_{jb} of macro-toryx associated with this star. *Black holes of our galaxy* are examples inverted stars.

Imaginary stars $(r_b < 0.5r_{jb})$ – In the imaginary stars, the outer radius r_b of a star body is less than one half of the eye radius r_{jb} of a macro-toryx associated with this star. *Supermassive black holes of quasars* are examples of imaginary stars.

12.9 Spacetime Limits of Stars

Section 8.3 describes the spacetime limits of micro-toryces making up elementary particles beyond which they do not interact with other elementary particles. Similarly, the spacetime limits of macro-toryces associated with celestial bodies correspond to the limits beyond which they do not interact with other celestial bodies.

Table 12.9. Maximum number of linear excitation quantum states n of macro-electrons associated with stars in various quantum spacetime levels of the Multiverse L.

Ratio of star mass m_b to solar mass m_s	Spacetime levels of the Multiverse			
	L0	*L1*	*L2*	*L3*
	$m = 1$	$m = 2$	$m = 3$	$m = 4$
1	$n = 6.418 \times 10^7$	$n = 175$	$n = 2$	$n = 0$
1000	$n = 6.418 \times 10^4$	$n = 0$	$n = 0$	$n = 0$
1000000	$n = 64$	$n = 0$	$n = 0$	$n = 0$
60000000	$n = 6$	$n = 0$	$n = 0$	$n = 0$

Table 12.9 shows the maximum number of linear excitation quantum states n of macro-electrons associated with stars in various quantum spacetime levels of the Multiverse L. These number of quantum states are defined based on the relative maximum and minimum relative spacetime radii b_p and b_p^{-1} expressed by Eqs. (8.2-1) and (8.2-2).

It follows from Table 12.9:

- The spacetime level of the Multiverse *L4* is the highest level within which the macro-electrons associated with the stars still exist at the linear excitation quantum state $n = 1$.
- As the spacetime level of the Multiverse decreases the maximum numbers of the linear excitation quantum states n increases and becomes the largest at the spacetime level of the Multiverse *L0*.

MAIN SUMMARY

- *Previous chapters described a role of the **micro-toryces** in the formation of matter particles and atoms of the micro-world. In the macro-world, the assemblies of atoms contained in each body form polarized **macro-toryces** that become integral parts of each body.*

- *Unified polarized macro-toryces produce **macro-trons** forming **toroidal celestial fields** associated with celestial bodies and responsible for interactions between them. Equations describing spacetime properties of micro- and macro-toryces are the same, except for the equations describing their eye radii.*

- *For large distances between celestial bodies, the **macro-toryx law of planetary motion** reduces to the Kepler's third law of planetary motion.*

- *Depending on the relationship between the outer radius of a star body and the eye radius of a macro-toryx associated with a star, the stars can be divided into three kinds: **outverted stars**, **inverted stars** and **imaginary stars.***

Notes:

<u>Notes:</u>

PART 3

Abstract & Applied
Mathematics
of a Helyx

13. HELYX BASIC STRUCTURE & PARAMETERS

CONTENTS

13.1 Helyx Basic Structure

Helyx is a dual-level helicola.

   The first helyx level – As shown in Fig. 13.1.1, the first helyx level is a three-dimensional (3D) single-level helicola. It is made up of a straight-line *leading string* with the radius $\tilde{r}_\infty$ and two branches m and n of a double-helical *trailing string* with the radius $\tilde{r}_1$. To visualize the double-helical trailing string, consider two points m and n rotating around a pivot point a with the rotational velocity $\tilde{V}_{1r}$, while the pivot point a propagates synchronously with the leading string along a straight line O_1O_1 with the translational velocity $\tilde{V}_{1t}$. The traces left by the moving points m and n form two branches of the double-helical trailing string of level 1: the trailing string m and the trailing string n.

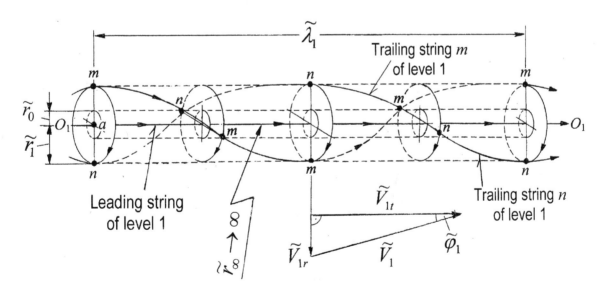

Figure 13.1.1. Structure of the first helyx level.

Consequently, the double-helical spiral propagates along its helical spiral paths m and n with the spiral velocity $\widetilde{V}_1$ that has two components, the translational velocity $\widetilde{V}_{1t}$ and the rotational velocity $\widetilde{V}_{1r}$, with all three velocities related to each other by the Pythagorean Theorem:

$$\widetilde{V}_1 = \sqrt{\widetilde{V}_{1t}^2 + \widetilde{V}_{1r}^2} \tag{13.1-1}$$

In the vector diagram of velocities shown in Fig. 13.1.1, $\widetilde{\varphi}_1$ is the **apex angle** of the first level of helyx leading string.

The second helyx level – As shown in Fig. 13.1.2, in the second helyx level, the double-helical trailing strings m and n of the first level with the radius $\widetilde{r}_1$ become the double-helical leading strings m and n of the second helyx level. The helyx double-helical trailing string has a cylindrical opening at the helyx center with the radius $\widetilde{r}_0$ called the **helyx eye**.

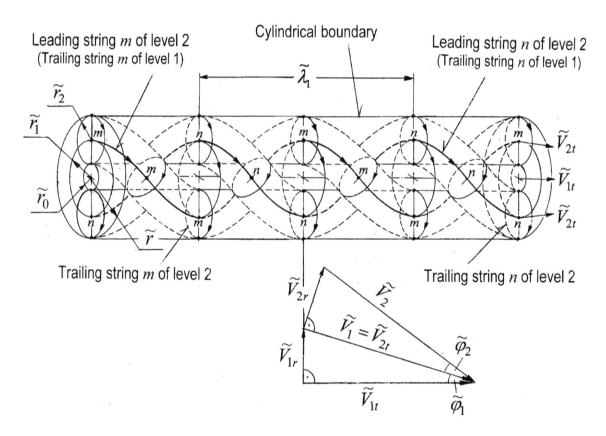

Figure 13.1.2. Structure of the first and second helyx levels.

A propagation of each helical spiral string m and n of the second helyx level is accompanied synchronously by a double helical trailing string with the radius $\widetilde{r}_3$. The helyx is contained within a cylindrical boundary with the radius that $\widetilde{r}$ is equal to:

$$\widetilde{r} = \widetilde{r_1} + \widetilde{r_2} \tag{13.1-2}$$

Figures 13.1.2 and 13.1.3 show a formation of the double helical trailing strings p_m and q_m associated with the helical trailing string m of the second helyx level. To visualize the double-helical trailing strings, consider two points, p and q, rotating around a pivot point m with the rotational velocity $\widetilde{V}_{2r}$, while the pivot point m propagates synchronously along the leading string m of the second level of helyx with the translational velocity $\widetilde{V}_{2t}$. The traces left by the moving points p_m and q_m form two branches of the double-helical trailing string of level 2: the trailing string p_m and the trailing string q_m.

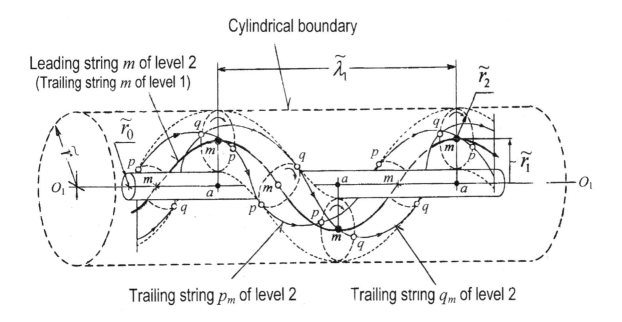

Figure 13.1.3. The double-helical leading strings m of the second helyx level accompanied by the strings p_m and q_m of the double-helical trailing string m.

Helical trailing strings of the helyx propagate along their helical paths in synchrony with the helyx leading strings, so, the translational velocity of the trailing string $\widetilde{V}_{2t}$ is equal to the spiral velocity of the helyx leading string $\widetilde{V}_1$:

$$\widetilde{V}_{2t} = \widetilde{V}_1 \tag{13.1-3}$$

As shown in Figure 13.1.2, the helyx velocities form sides of right triangles. Therefore, it is possible establish the relationships between these parameters by using the Pythagorean Theorem:

$$\tilde{V}_2 = \sqrt{\tilde{V}_{2t}^2 + \tilde{V}_{2r}^2}$$

(13.1-4)

Helyx spin - Helyx may have either up or down spins, with both of them defined by the right-hand rule as shown in Fig. 13.1.4. The helyx spin depends on the directions of the rotational velocities of helyx trailing string $\tilde{V}_{2r}$.

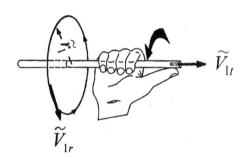

Figure 13.1.4. The up helyx spin.

13.1.2 Helyx Spacetime Parameters in Absolute Units

$\tilde{f}_0$ = helyx base frequency

$\tilde{f}_1$ = frequency of the helyx level 1

$\tilde{f}_2$ = frequency of the helyx level 2

$\tilde{L}_1$ = spiral length the helyx level 1

$\tilde{L}_2$ = spiral length the helyx level 1

$\tilde{r}$ = radius of helyx cylindrical boundary

$\tilde{r}_0$ = helyx eye radius

$\tilde{r}_1$ = radius of trailing string of the helyx level 1 = radius of leading string of the helyx level 2

$\tilde{r}_2$ = radius of trailing string of the helyx level 2

$\tilde{T}_0$ = helyx base period

$\tilde{T}_1$ = period the helyx level 1

$\tilde{T}_2$ = period of of the helyx level 2

$\tilde{V}_1$ = spiral velocity the helyx level 1

$\tilde{V}_{1r}$ = rotational velocity of helyx level 1

$\tilde{V}_{1t}$ = translational velocity the helyx level 1

$\tilde{V}_2$ = spiral velocity of of the helyx level 2

$\tilde{V}_{2r}$ = rotational velocity of the helyx level 2

$\widetilde{V}_{2t}$ = translational velocity of the helyx level 2

$\widetilde{\lambda}_1$ = wavelength the helyx level 1

$\widetilde{\lambda}_2$ = wavelength of the helyx level 2

$\widetilde{\varphi}_1$ = apex angle the helyx level 1

$\widetilde{\varphi}_2$ = apex angle of the helyx level 2.

In the hodograph shown in Figure 13.1.2, the relative spacetime parameters of the helyx level 1 and level 2 form sides of right triangles. These parameters are related to the middle points of helyx trailing strings.

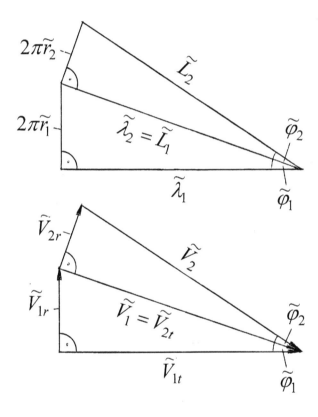

Figure 13.2. Hodographs of the helyx absolute spacetime parameters related to the middle points of helyx trailing strings.

13.3 Helyx Basic Genetic Codes in Absolute Units

Notably, the symbols used for defining the helyx parameters are the same as those used for the toryx, except for the "wave" mark (tilde) over the symbols of the helyx parameters. Below is a list of the helyx spacetime parameters expressed in absolute values.

The helyx basic genetic codes include three fundamental equations limiting the degrees of freedom of several helyx parameters (see Exhibit 13.3). These genetic codes provide the simplest way to derive the relationships between all spacetime parameters of helyx. In spite of their outmost simplicity, these genetic codes provide helyces with amazing spacetime properties, including a capability to exist in four unique topologically-polarized states within the range of the radius of helyx leading strings r_1 extending from negative to positive infinity $(-\infty < r_1 < +\infty)$ as will be described in Chapter 5.

Exhibit 13.3. Helyx basic genetic codes in absolute units.

- The frequencies of the first and second levels of helyx trailing string $\tilde{f}_1$ and $\tilde{f}_2$ are equal to one another:

$$\tilde{f}_1 = \tilde{f}_2 \tag{13.3-1}$$

- The helyx eye radius $\tilde{r}_0$ is equal to the toryx eye radius r_0 that is constant:

$$\tilde{r}_0 = r_0 = \tilde{r}_1 - \tilde{r}_2 = const. \tag{13.3-2}$$

- The spiral velocity $\tilde{V}_2$ of the second level helyx trailing string is constant and equals to the velocity of light c at each point of its spiral path:

$$\tilde{V}_2 = \sqrt{\tilde{V}_{2t}^2 + \tilde{V}_{2r}^2} = c = const. \tag{13.3-3}$$

Since the helyx eye radius $\tilde{r}_0$ is equal to the toryx eye radius r_0, the helyx base frequency $\tilde{f}_0$ is equal to the toryx base frequency f_0 and the helyx base period $\tilde{T}_0$ is equal to the toryx base period T_0.

$$\tilde{f}_0 = f_0 = \frac{c}{2\pi\tilde{r}_0} = \frac{c}{2\pi r_0} \tag{13.3-4}$$

$$\tilde{T}_0 = T_0 = \frac{2\pi\tilde{r}_0}{c} = \frac{2\pi r_0}{c} \tag{13.3-5}$$

13.4 Helyx Basic Genetic Codes in Relative Units

The helyx basic genetic codes can be simplified by expressing the helyx spacetime parameters in relative units in respect to the helyx eye radius $\tilde{r}_0$, the velocity of light c and the helyx base frequency $\tilde{f}_0$ as shown in Table 13.4.

Table 13.4. Helyx relative spacetime parameters.

Helyx relative parameters	Equations
Radius of helyx cylindrical boundary	$\tilde{b} = \tilde{r} / \tilde{r}_0$
Radius of trailing string of the helyx level 1 Radius of leading string of the helyx level 2	$\tilde{b}_1 = \tilde{r}_1 / \tilde{r}_0$
Radius of trailing string of the helyx level 2	$\tilde{b}_2 = \tilde{r}_2 / \tilde{r}_0$
Spiral length of the helyx level 1	$\tilde{l}_1 = \tilde{L}_1 / 2\pi\tilde{r}_0$
Spiral length of the helyx level 2	$\tilde{l}_2 = \tilde{L}_2 / 2\pi\tilde{r}_0$
Period of helyx the level 1	$\tilde{t}_1 = \tilde{T}_1 / \tilde{T}_0$
Period of helyx the level 2	$\tilde{t}_2 = \tilde{T}_2 / \tilde{T}_0$
Spiral velocity of the helyx level 1	$\tilde{\beta}_1 = \tilde{V}_1 / c$
Translational velocity of the helyx level 1	$\tilde{\beta}_{1t} = \tilde{V}_{1t} / c$
Rotational velocity of the helyx level 1	$\tilde{\beta}_{1r} = \tilde{V}_{1r} / c$
Spiral velocity of the helyx level 2	$\tilde{\beta}_2 = \tilde{V}_2 / c$
Translational velocity of the helyx level 2	$\tilde{\beta}_{2t} = \tilde{V}_{2t} / c$
Rotational velocity of the helyx level 2	$\tilde{\beta}_{2r} = \tilde{V}_{2r} / c$
Frequency of the helyx level 1	$\tilde{\delta}_1 = \tilde{f}_1 / \tilde{f}_0$
Frequency of the helyx level 2	$\tilde{\delta}_2 = \tilde{f}_2 / \tilde{f}_0$
Wavelength the helyx level 1	$\tilde{\eta}_1 = \tilde{\lambda}_1 / 2\pi\tilde{r}_0$
Wavelength of the helyx level 2	$\tilde{\eta}_2 = \tilde{\lambda}_2 / 2\pi\tilde{r}_0$

13.5 Helyx Basic Genetic Codes in Relative Units

Exhibit 13.5 shows three helyx basic genetic codes in relative units.

Exhibit 13.5. Helyx basic genetic codes in relative units.

- The relative frequencies of the first and second levels of helyx trailing string $\tilde{\delta}_1$ and $\tilde{\delta}_2$ are equal to one another:

$$\tilde{\delta}_1 = \tilde{\delta}_2 \tag{13.5-1}$$

- The difference between the relative radius of helyx leading string $\tilde{b}_1$ and the relative radius of helyx trailing string $\tilde{b}_2$ is equal to 1:

$$\tilde{b}_1 - \tilde{b}_2 = 1 \tag{13.5-2}$$

- The relative spiral velocity $\tilde{\beta}_2$ of the helyx level 2 is equal to 1 at each point of its spiral path:

$$\tilde{\beta}_2 = \sqrt{\tilde{\beta}_{2t}^2 + \tilde{\beta}_{2r}^2} = 1 \tag{13.5-3}$$

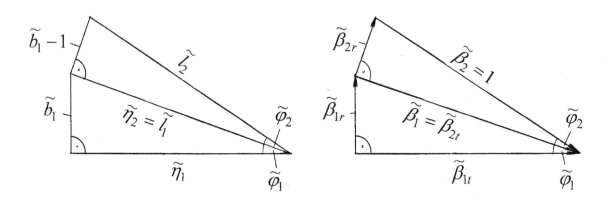

Figure 13.5. Hodographs of the helyx relative spacetime parameters related to the middle points of helyx trailing strings.

In the hodograph shown in Figure 13.5, the relative spacetime parameters of the helyx level 1 and level 2 form sides of right triangles. These parameters are related to the middle points of helyx trailing strings.

13.6 Derivative Spacetime Equations of Helyces

Based on the structure of the basic helyx and also on equations of the helyx basic genetic codes, it is possible to derive equations for all other helyx spacetime parameters shown in Tables 13.6.1 and 13.6.2.

Table 13.6.1. Spacetime parameters of the helyx level 1 and 2 as the functions of the relative radius of the helyx leading string $\tilde{b}_1$.

Relative parameter	Helyx level 1 Eq.(a)	Helyx level 2 Eq. (b)
Radius Eq. (13.6-1)	$\tilde{b}_1 = \dfrac{\tilde{r}_1}{\tilde{r}_0}$	$\tilde{b}_2 = \tilde{b}_1 - 1$
Apex angle Eq. (13.6-2)	$\sin s\tilde{\varphi}_1 = \dfrac{\tilde{b}_1\sqrt{2\tilde{b}_1 - 1}}{(\tilde{b}_1 - 1)^2}$	$\cos s\tilde{\varphi}_2 = \dfrac{\tilde{b}_1 - 1}{\tilde{b}_1}$
Wavelength Eq. (13.6-3)	$\tilde{\eta}_1 = \dfrac{\tilde{\lambda}_1}{2\pi r_0} = \sqrt{\dfrac{(\tilde{b}_1 - 1)^4 - \tilde{b}_1^2(2\tilde{b}_1 - 1)}{2\tilde{b}_1 - 1}}$	$\tilde{\eta}_2 = \dfrac{\tilde{\lambda}_2}{2\pi r_0} = \dfrac{(\tilde{b}_1 - 1)^2}{\sqrt{2\tilde{b}_1 - 1}}$
Length of one winding Eq. (13.6-4)	$\tilde{l}_1 = \dfrac{\tilde{L}_1}{2\pi r_0} = \dfrac{(\tilde{b}_1 - 1)^2}{\sqrt{2\tilde{b}_1 - 1}}$	$\tilde{l}_2 = \dfrac{\tilde{L}_2}{2\pi r_0} = \dfrac{\tilde{b}_1(\tilde{b}_1 - 1)}{\sqrt{2\tilde{b}_1 - 1}}$
Translational velocity Eq. (13.6-5)	$\tilde{\beta}_{1t} = \dfrac{\tilde{V}_{1t}}{c} = \dfrac{\sqrt{(\tilde{b}_1 - 1)^4 - \tilde{b}_1^2(2\tilde{b}_1 - 1)}}{\tilde{b}(\tilde{b}_1 - 1)}$	$\tilde{\beta}_{2t} = \dfrac{\tilde{V}_{2t}}{c} = \dfrac{\tilde{b}_1 - 1}{\tilde{b}_1}$
Rotational velocity Eq. (13.6-6)	$\tilde{\beta}_{1r} = \dfrac{\tilde{V}_{1r}}{c} = \dfrac{\sqrt{2\tilde{b}_1 - 1}}{\tilde{b}_1 - 1}$	$\tilde{\beta}_{2r} = \dfrac{\tilde{V}_{2r}}{c} = \dfrac{\sqrt{2\tilde{b}_1 - 1}}{\tilde{b}_1}$
Spiral velocity Eq. (13.6-7)	$\tilde{\beta}_1 = \dfrac{\tilde{V}_1}{c} = \dfrac{\tilde{b}_1 - 1}{\tilde{b}_1}$	$\tilde{\beta}_2 = \dfrac{\tilde{V}_2}{c} = 1$
Frequency Eq. (13.6-8)	$\tilde{\delta}_1 = \tilde{\delta}_2 = \tilde{\delta} = \dfrac{\tilde{f}_1}{\tilde{f}_0} = \dfrac{\tilde{f}_2}{\tilde{f}_0} = \dfrac{\sqrt{2\tilde{b}_1 - 1}}{\tilde{b}_1(\tilde{b}_1 - 1)}$	
Period Eq. (13.6-9)	$\tilde{\tau}_1 = \tilde{\tau}_2 = \tau = \dfrac{\tilde{T}_1}{\tilde{T}_0} = \dfrac{\tilde{T}_2}{\tilde{T}_0} = \dfrac{\tilde{b}_1(\tilde{b}_1 - 1)}{\sqrt{2\tilde{b}_1 - 1}}$	

The relative radius of cylindrical boundary is equal to:

$$\tilde{b} = 2\tilde{b}_1 - 1 \qquad (13.6\text{-}10)$$

The helyx trigonometric function $\cos s\tilde{\varphi}_2$ in Eq. (13.6-2b) relates to the elementary trigonometric function $\cos\varphi_2$ as follows:

$$\cos s\tilde{\varphi}_2 = \cos\tilde{\varphi}_2 \quad (0 < \varphi_2 < 180^0) \qquad (13.6\text{-}11)$$

$$\cos s\tilde{\varphi}_2 = 1/\cos\tilde{\varphi}_2 \quad (180^0 < \varphi_2 < 360^0) \qquad (13.6\text{-}12)$$

Table 13.6.2. Spacetime parameters of the helyx levels 1 and 2 as the functions of the apex angle $\tilde{\varphi}_2$ of the helyx level 2.

Relative parameter	Helyx level 1 Eq.(a)	Helyx level 2 Eq. (b)
Radius Eq. (13.6-13)	$\tilde{b}_1 = \dfrac{1}{1 - \cos s\tilde{\varphi}_2}$	$\tilde{b}_2 = \dfrac{\cos s\tilde{\varphi}_2}{1 - \cos s\tilde{\varphi}_2}$
Wavelength Eq. (13.6-14)	$\tilde{\eta}_1 = \dfrac{\sqrt{\cos s^2\tilde{\varphi}_2 - \tan s^2\tilde{\varphi}_2}}{\tan s\tilde{\varphi}_2 - \sin s\tilde{\varphi}_2}$	$\eta_2 = \dfrac{\cos s\tilde{\varphi}_2}{(1 - \cos s\tilde{\varphi}_2)\tan s\tilde{\varphi}_2}$
Length (Eq. 13.6-15)	$\tilde{l}_1 = \dfrac{\cos s\tilde{\varphi}_2}{(1 - \cos s\tilde{\varphi}_2)\tan s\tilde{\varphi}_2}$	$\tilde{l}_2 = \dfrac{1}{\tan s\tilde{\varphi}_2 - \sin s\tilde{\varphi}_2}$
Translational velocity (Eq. 13.6-16)	$\tilde{\beta}_{1t} = \sqrt{\cos s^2\tilde{\varphi}_2 - \tan s^2\tilde{\varphi}_2}$	$\tilde{\beta}_{2t} = \cos s\tilde{\varphi}_2$
Rotational velocity Eq. (13.6-17)	$\tilde{\beta}_{1r} = \tan s\tilde{\varphi}_2$	$\tilde{\beta}_{2r} = \sin s\tilde{\varphi}_2$
Spiral velocity Eq. (13.6-18)	$\tilde{\beta}_1 = \cos s\tilde{\varphi}_2$	$\tilde{\beta}_2 = 1$
Frequency Eq. (13.6-19)	$\tilde{\delta}_1 = \tilde{\delta}_2 = \tilde{\delta} = \tan s\tilde{\varphi}_2 - \sin s\tilde{\varphi}_2$	
Period Eq. (13.6-20)	$\tilde{\tau}_1 = \tilde{\tau}_2 = \tilde{\tau} = \dfrac{1}{\tan s\tilde{\varphi}_2 - \sin s\tilde{\varphi}_2}$	

The relative radius of cylindrical boundary $\tilde{b}$ is equal to:

$$\tilde{b} = \frac{\tilde{r}}{\tilde{r}_0} = \frac{1 + \cos s\tilde{\varphi}_2}{1 - \cos s\tilde{\varphi}_2} \qquad (13.6\text{-}21)$$

The relative spiral length of trailing string $\widetilde{l_2}$ is equal to:

$$\widetilde{l_2} = \frac{1}{\widetilde{\delta}} = \widetilde{\tau} \qquad (13.6\text{-}22)$$

MAIN SUMMARY

- *Helyx is a particular case of a dual-level helicola (Figs 13.1.1 - 13.1.3). Its first level is made up of a straight-line leading string and a double-helical trailing string. The double-helical trailing string of the first level becomes a leading string of the helyx second level. Its each branch is accompanied by the double-helical trailing string of the helyx second level.*

- *Main spacetime properties of a helyx are based on **three basic helyx genetic codes** (Exhibits 13.3 & 13.5) limiting its degree of freedom, so it becomes possible to establish a relationship between helyx spacetime parameters (Tables 13.6.1 and 13.6.2).*

14. FEATURES OF ABSTRACT MATHEMATICS OF A HELYX

Equations describing the helyx spacetime parameters are mostly based on elementary math commonly taught in high schools. However, to satisfy the helyx spacetime postulates, it is necessary to modify several aspects of elementary math, including the definitions of zero, number line and elementary trigonometric functions. Also, unlike the elementary math that deals with stationary spiral elements, the helyx math considers the spiral elements in motion.

14.1 Infinility versus Elementary Zero

Conventionally, we use the elementary zero (0) in two ways. Firstly, we use it for counting of non-divisible entities. In an elementary number line (Fig. 14.1.1) it appears as an integer immediately preceding number one (1).

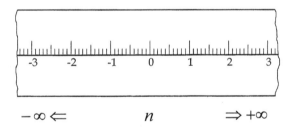

Figure 14.1.1. Elementary number line.

Secondly, we use zero to represent the absolute absence of any quantity and quality. Mathematically, the elementary zero (0) is equal to a ratio of one (1) to infinity (∞). The helyx math clearly separates two applications of zero described above. The zero is still considered as an integer for counting of non-divisible entities and still retains its old symbol (0). But, in application to the spacetime entities the zero is replaced with a quantity that is infinitely approaching to it. This quantity is called **infinility**, from the "infinite nil." (Notably, the term infinility is used in the helyx math instead of the known math term **infinitesimal**). In the helyx math, both infinity and infinility can be positive, negative, real and imaginary as shown below.

$$\text{Real infinity: } \pm\infty = \frac{1}{\pm 0}; \quad \text{Imaginary infinity: } \pm\infty i = \frac{1}{\pm 0i}$$

$$\text{Real infinility: } \pm 0 = \frac{1}{\pm\infty}; \quad \text{Imaginary infinility: } \pm 0i = \frac{1}{\pm\infty i}$$

Figure 14.1.2 shows symbolically positive and negative infinities $(\pm\infty)$ and also positive and negative infinility (± 0) as equal counterparts in respect to the positive and negative unities (± 1).

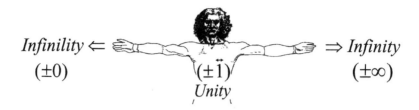

Infinility $\Leftarrow$ $\Rightarrow$ *Infinity*
   (± 0)   ($\pm \vec{1}$)   ($\pm \infty$)
                               Unity

Figure 14.1.2. Infinity ($\pm \infty$), infinility (± 0) and unity (± 1).

14.2 Helyx Spacetime Trigonometry

Definitions of elementary trigonometric functions are based on transformations of a right triangle as a function of the non-right angle $\widetilde{\varphi}_2$ (Fig. 14.2.1):

$$\cos \varphi_2 = x \qquad (0^0 < \varphi_2 < 360^0) \tag{14.2-1}$$

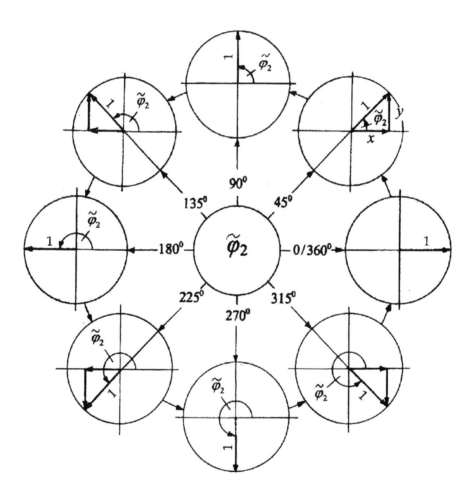

Figure 14.2.1. Transformations of a right triangle in elementary trigonometry.

The main features of the transformations shown in Fig. 14.2.1 are:

- When the length of the hypotenuse of the triangles is equal to 1, the ranges of the lengths of its sides x and y are between 1 and -1.
- The triangles located in two left quadrants are the mirror images of the triangles located in two right quadrants.
- The triangles located in two bottom quadrants are the mirror images of the triangles located at two top quadrants.

In the spacetime trigonometry, the transformations of the right triangle are partially modified to satisfy the helyx spacetime postulates. Consequently, the helyx spacetime trigonometric function $\cos s\widetilde{\varphi}_2$ relates to the elementary trigonometric function $\cos\widetilde{\varphi}_2$ as follows:

$$\cos s\widetilde{\varphi}_2 = \cos\widetilde{\varphi}_2 \quad (0 < \widetilde{\varphi}_2 < 180^0) \tag{14.2-2}$$

$$\cos s\widetilde{\varphi}_2 = 1/\cos\widetilde{\varphi}_2 \quad (180^0 < \widetilde{\varphi}_2 < 360^0) \tag{14.2-3}$$

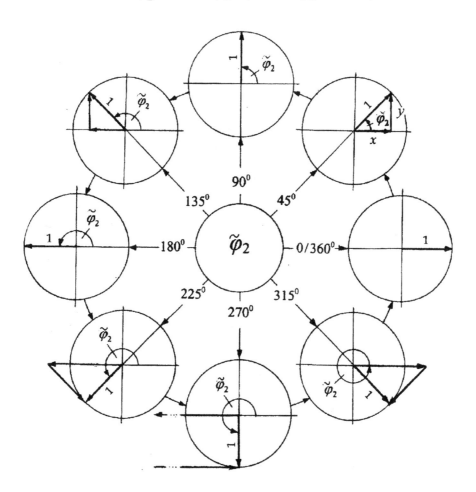

Figure 14.2.2. Transformations of a right triangle
in the spacetime trigonometry.

The main features of the transformations shown in Fig. 14.2.2 are:

- When the angle $\widetilde{\varphi}_2$ is between 0 and 180^0, the right triangles are the same as in elementary trigonometry. Thus, within this range of the angle $\widetilde{\varphi}_2$ the elementary and spacetime trigonometry are based on the same principle.
- When the angle $\widetilde{\varphi}_2$ is between 180 and 360^0, the right triangle becomes **outverted**. Consequently, the length of its horizontal side x becomes greater than 1, while the length of the other side y is expressed with imaginary numbers.
- When the angle $\widetilde{\varphi}_2$ approaches 270^0 from the angle smaller than 270^0, the length of its horizontal side x approaches real positive infinity $(+\infty)$, while the length of the other side y approaches imaginary positive infinity $(+\infty i)$.
- When the angle $\widetilde{\varphi}_2$ approaches 270^0 from the angle greater than 270^0, the length of its horizontal side x approaches real negative infinity $(-\infty)$, while the length of the other side y approaches imaginary negative infinity $(-\infty i)$.
- When the angle $\widetilde{\varphi}_2$ approaches 360^0 from the angle smaller than 360^0, the length of its horizontal side x approaches 1, while the length of the other imaginary side y approaches imaginary negative infinility $(-0i)$.

14.3 Helyx Number Lines

We consider below three kinds of helyx number lines that are directly related to the helyx parameters:

- ***Helyx vorticity*** $\widetilde{V}$ **number line**
- ***Helyx reality*** $\widetilde{R}$ **number line**
- ***Helyx boundary*** $\widetilde{B}$ **number line.**

Both kinds of number lines are presented below in the forms of circular diagrams in which the numbers $\widetilde{V}$, $\widetilde{R}$ and $\widetilde{B}$ are expressed as functions of the steepness angle of helyx trailing string $\widetilde{\varphi}_2$.

Helyx vorticity $\widetilde{V}$ **number line** - In the helyx vorticity $\widetilde{V}$ number line (Fig. 14.3.1), the real numbers $\widetilde{V}$ are equal to the ratio of the radius of helyx trailing string $\widetilde{b}_2$ to the radius of helyx leading string $\widetilde{b}_1$ with an opposite sign. These numbers are extended clockwise along a circle from the real positive infinity $(+\infty)$ to the real negative infinity $(-\infty)$ as a function of the apex angle of trailing string $\widetilde{\varphi}_2$.

$$\widetilde{V} = -\frac{\widetilde{b}_2}{\widetilde{b}_1} = -\cos s\widetilde{\varphi}_2 \qquad (14.3\text{-}1)$$

The helyx vorticity $\widetilde{V}$ number line is divided into two domains, the $\widetilde{V}$ *infinility domain* and the $\widetilde{V}$ *infinity domain*, occupying equal sectors of the circular number line.

- The $\widetilde{V}$ infinility domain occupies two top quadrants; it contains the values of $\widetilde{V}$ extending clockwise from the real positive unity $(+1)$ and passing through infinility (± 0) to the real negative unity (-1).

- The $\widetilde{V}$ infinity domain resides in two bottom quadrants; it contains the values of $\widetilde{V}$ extending counterclockwise from the real positive unity $(+1)$ and passing through infinity $(\pm\infty)$ to the real negative unity (-1).

In the helyx vorticity $\widetilde{V}$ number line, the real positive infinility $(+0)$ merges with real negative infinility (-0) at $\widetilde{\varphi}_2 = 90^0$, while real negative infinity $(-\infty)$ merges with real positive infinity $(+\infty)$ at $\widetilde{\varphi}_2 = 270^0$.

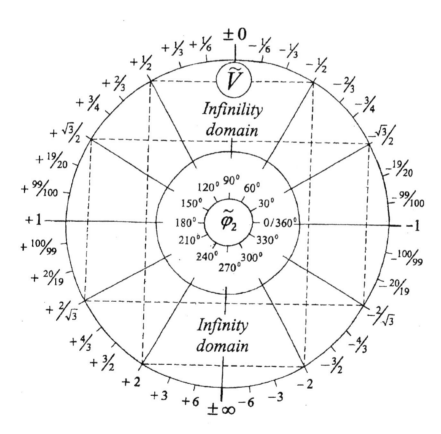

Figure 14.3.1. The helyx vorticity $\widetilde{V}$ number line.

Symmetries between the helyx vorticity $\widetilde{V}$ numbers - There are two kinds of symmetries between the numbers $\widetilde{V}$ that belong to the four quadrants of circular diagram, the *inverse $\widetilde{V}$ -symmetry* and the *reverse $\widetilde{V}$ -symmetry*.

- In the inverse $\widetilde{V}$-symmetry, the magnitudes of the numbers $\widetilde{V}$ located in the top quadrants are inversed (reciprocated) in respect to the magnitudes of the numbers $\widetilde{V}$ located in the bottom quadrants.
- In the reverse $\widetilde{V}$-symmetry, the numbers $\widetilde{V}$ located in the right quadrants and the left quadrants have the same magnitudes but reversed signs.

Helyx reality $\widetilde{R}$ number line - In the helyx reality $\widetilde{R}$ number line (Fig. 14.3.2), the real and imaginary numbers $\widetilde{R}$ are equal to the ratio of helyx trailing string $\widetilde{r}_2$ to the radius of helyx eye $\widetilde{r}_0$.

$$\widetilde{R} = \frac{\widetilde{r}_2}{\widetilde{r}_0} = \sqrt{\frac{1 + \cos s\widetilde{\varphi}_2}{1 - \cos s\widetilde{\varphi}_2}} \tag{14.3-2}$$

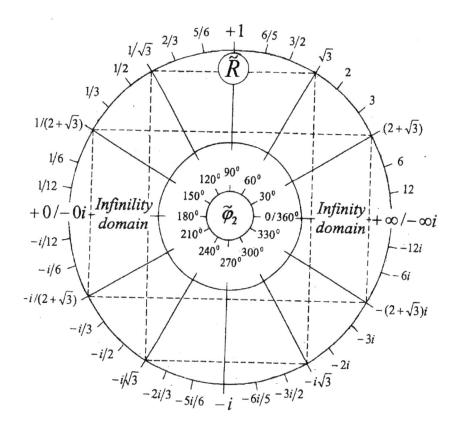

Figure 14.3.2. The helyx reality $\widetilde{R}$ number line.

These numbers are extended counterclockwise along a circle from the real positive infinity $(+\infty)$ to the imaginary negative infinity $(-\infty i)$ as a function of the steepness angle of trailing string $\widetilde{\varphi}_2$.

The helyx reality $\widetilde{R}$ number line is divided into two domains, the $\widetilde{R}$ *infinility domain* and the $\widetilde{R}$ *infinity domain*, occupying equal sectors of the circular number line.

- The $\widetilde{R}$ <u>infinility domain</u> occupies two left quadrants; it contains the values of $\widetilde{R}$ extending counterclockwise from the real positive unity $(+1)$ and passing through infinility $(+0/0i)$ to the imaginary negative unity $(-i)$.
- The $\widetilde{R}$ <u>infinity domain</u> resides in two right quadrants; it contains the values of $\widetilde{R}$ extending clockwise from the real positive unity $(+1)$ and passing through real positive and imaginary negative infinities $(+\infty/-\infty i)$ to the imaginary negative unity $(-i)$.

In the helyx reality $\widetilde{R}$ number line, the real positive infinility $(+0)$ merges with the imaginary negative infinility $(-0i)$ at $\widetilde{\varphi}_2 = 180^0$, while real positive infinity $(+\infty)$ merges with imaginary negative infinity $(-\infty i)$ at $\widetilde{\varphi}_2 = 360^0$.

Symmetries between the helyx realities $\widetilde{R}$ numbers - There are two kinds of symmetries between the numbers $\widetilde{R}$ that belong to the four quadrants of circular diagram, the *inverse $\widetilde{R}$-symmetry* and the *reverse reality $\widetilde{R}$-symmetry*.

- In the inverse $\widetilde{R}$-symmetry, the magnitudes of the numbers $\widetilde{R}$ located in the left quadrants are inversed (reciprocated) in respect to the magnitudes of the numbers $\widetilde{R}$ located in the right quadrants.
- In the reverse reality $\widetilde{R}$-symmetry, the numbers $\widetilde{R}$ located in the top quadrants are real positive, while these numbers in the bottom quadrants are imaginary negative.

Helyx boundary $\widetilde{B}$ number line - In the helyx boundary $\widetilde{B}$ number line (Fig. 14.3.3), the real numbers $\widetilde{B}$ are equal to the ratio of the radius of helyx radius of spherical boundary $\widetilde{r}$ to the helyx eye radius $\widetilde{r}_0$.

$$\widetilde{B} = \frac{\widetilde{r}}{\widetilde{r}_0} = \frac{1 + \cos s\widetilde{\varphi}_2}{1 - \cos s\widetilde{\varphi}_2} \tag{14.3-3}$$

These numbers are extended counterclockwise along a circle from the real positive infinity $(+\infty)$ to the real negative infinity $(-\infty)$ as a function of the apex angle of trailing string $\widetilde{\varphi}_2$.

The helyx boundary $\widetilde{B}$ number line is divided into two domains, the $\widetilde{B}$ *infinility domain* and the $\widetilde{B}$ *infinity domain*, occupying equal sectors of the circular number line.

- The $\widetilde{B}$ <u>infinility domain</u> occupies two left quadrants; it contains the values of $\widetilde{B}$ extending counterclockwise from the real positive unity $(+1)$ and passing through infinility $(+0/-0)$ to the real negative unity (-1).

- The $\widetilde{B}$ infinity domain resides in two right quadrants; it contains the values of $\widetilde{B}$ extending clockwise from the real positive unity $(+1)$ and passing through real positive and negative infinities $(+\infty/-\infty)$ to the real negative unity (-1).

In the helyx boundary $\widetilde{B}$ number line, real positive infinility $(+0)$ merges with real negative infinility (-0) at $\widetilde{\varphi}_2 = 180^0$, while real positive infinity $(+\infty)$ merges with real negative infinity $(-\infty)$ at $\widetilde{\varphi}_2 = 360^0$.

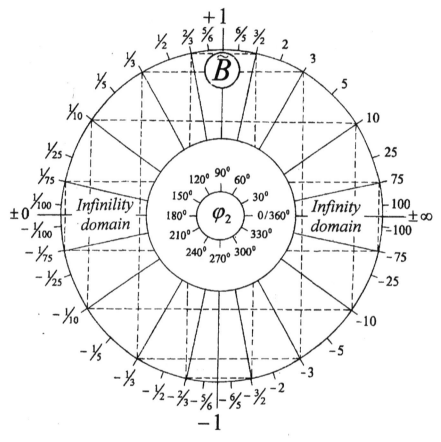

Figure 14.3.3. The helyx boundary $\widetilde{B}$ number line.

Symmetries between the helyx realities $\widetilde{B}$ numbers - There are two kinds of symmetries between the numbers $\widetilde{B}$ that belong to the four quadrants of circular diagram, the *inverse $\widetilde{B}$ - symmetry* and the *reverse $\widetilde{B}$ - symmetry*.

- In the inverse $\widetilde{B}$ -symmetry, the magnitudes of the numbers $\widetilde{B}$ located in the left quadrants are inversed (reciprocated) in respect to the magnitudes of the numbers $\widetilde{B}$ located in the right quadrants.

- In the reverse $\widetilde{B}$ - symmetry, the numbers $\widetilde{B}$ located in the top and bottom quadrants have the same magnitudes but reversed signs.

14.4. Helyx Parameters in Number Lines & Trigonometry

The numbers $\widetilde{V}$ of the helyx vorticity $\widetilde{V}$ number line (Fig. 14.3.1) relate to the helyx parameters by the equation:

$$\widetilde{V} = -\frac{\widetilde{r}_2}{\widetilde{r}_1} = -\frac{\widetilde{b}_2}{\widetilde{b}_1} = -\frac{\widetilde{b}_1 - 1}{\widetilde{b}_1} = -\frac{\widetilde{b} - 1}{\widetilde{b} + 1} = -\widetilde{\beta}_{2t} = -\cos s\widetilde{\varphi}_2 \tag{14.4-1}$$

The numbers $\widetilde{R}$ of the helyx reality $\widetilde{R}$ number line (Fig. 14.3.2) relate to the helyx parameters by the equation:

$$\widetilde{R} = \sqrt{\widetilde{b}} = \sqrt{2\widetilde{b}_1 - 1} = \sqrt{\frac{1 + \cos s\widetilde{\varphi}_2}{1 - \cos s\widetilde{\varphi}_2}} \tag{14.4-2}$$

The numbers $\widetilde{B}$ of the helyx boundary $\widetilde{B}$ number line (Fig. 14.3.3) relate to the helyx parameters by the equation:

$$\widetilde{B} = \widetilde{b} = \frac{\widetilde{r}}{\widetilde{r}_0} = 2\widetilde{b}_1 - 1 = \frac{1 + \cos s\widetilde{\varphi}_2}{1 - \cos s\widetilde{\varphi}_2} \tag{14.4-3}$$

Figure 14.4.3 shows the application of the spiral spacetime math for the calculation of relative velocities of helyx trailing string as its steepness angle $\widetilde{\varphi}_2$ increases from 0 to 360^0. In each right triangle of velocities of trailing string one side represents the relative translational velocity $\widetilde{\beta}_{2t}$ and the other side the relative rotational velocity $\widetilde{\beta}_{2r}$, while its hypotenuse represents the relative spiral velocity $\widetilde{\beta}_2 = 1$. There is a clear similarity between the transformations of the velocities of the helyx trailing string shown in Figure 14.3.3 and the transformations of a right triangle corresponding to the helyx trigonometry shown in Figure 14.2.2.

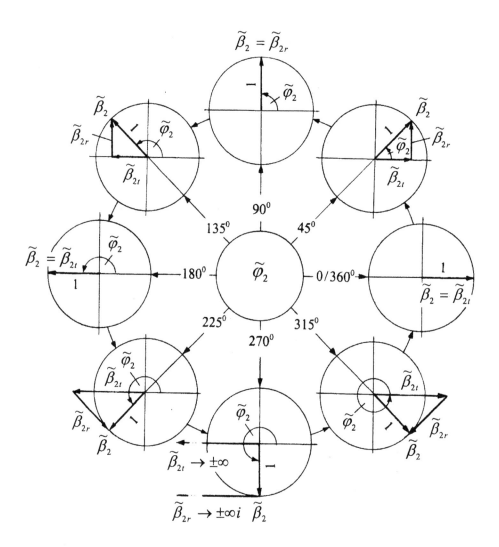

Figure 14.4.3. Transformations of right triangle representing vectors of the relative velocities of helyx trailing string $\tilde{\beta}_2$, $\tilde{\beta}_{2t}$ and $\tilde{\beta}_{2r}$.

MAIN SUMMARY

Based on three helyx spacetime postulates, it is necessary to modify three commonly-accepted aspects of elementary mathematics.

- *Conventional zero is replaced with **infinility** (± 0) that is an inverse of **infinity** $(\pm \infty)$.*

- *Trigonometric **cosine** function between 180^0 and 360^0 is replaced with its inverse value.*

- *Three **circular number lines** are used to provide a possibility to express a full range of toryx parameters and to reveal symmetrical relationships between them.*

Notes

15. CLASSIFICATION OF HELYCES

CONTENTS

15.1 Main Groups & Subgroups of Helyces

Helyces are divided into four main groups and eight subgroups according to their vorticity $\widetilde{V}$ and reality $\widetilde{R}$ as shown in Tables 15.1.1, 15.1.2 and Figs. 15.1.1, 15.1.2.

Table 15.1.1. Realities $\widetilde{R}$ and vorticities $\widetilde{V}$ of helyces of main groups.

Helyx name	$\widetilde{\varphi}_2$	$\widetilde{R}$	$\widetilde{V}$
Real negative	0^0 - 90^0	Real	$(-)$
Real positive	90^0 - 180^0	Real	$(+)$
Imaginary positive	180^0 - 270^0	Imaginary	$(+)$
Imaginary negative	270^0 - 360^0	Imaginary	$(-)$

Real negative helyces $(0^0 < \widetilde{\varphi}_2 < 90^0)$ – The real negative helyces are located in the top right quadrants of the circular diagrams. In these helyces, the helyx reality $\widetilde{R}$ is expressed with real numbers and the helyx vorticity $\widetilde{V}$ with negative numbers.

Real positive helyces $(90^0 < \widetilde{\varphi}_2 < 180^0)$ – The real positive helyces are located in the top left quadrants of the circular diagrams. In these helyces, the helyx reality $\widetilde{R}$ is expressed with real numbers and the helyx vorticity $\widetilde{V}$ with positive numbers.

Imaginary positive helyces $(180^0 < \widetilde{\varphi}_2 < 270^0)$ – The imaginary positive helyces are located in the bottom left quadrants of the circular diagrams. In these helyces, the helyx reality $\widetilde{R}$ is expressed with imaginary numbers and the helyx vorticity $\widetilde{V}$ with positive numbers.

Imaginary negative helyces $(270^0 < \widetilde{\varphi}_2 < 360^0)$ – The imaginary negative helyces are located in the bottom right quadrants of the circular diagrams. In these helyces, the helyx reality $\widetilde{R}$ is expressed with imaginary numbers and the helyx vorticity $\widetilde{V}$ with negative numbers.

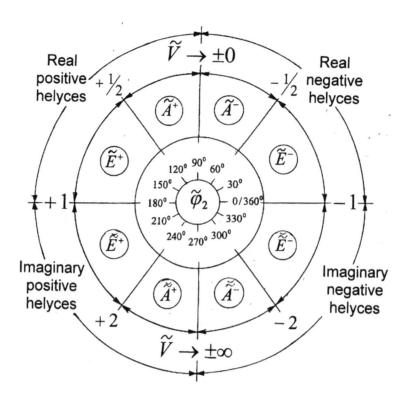

Figure 15.1.1. Main groups and subgroups of helyces as a function of helyx vorticity $\widetilde{V}$.

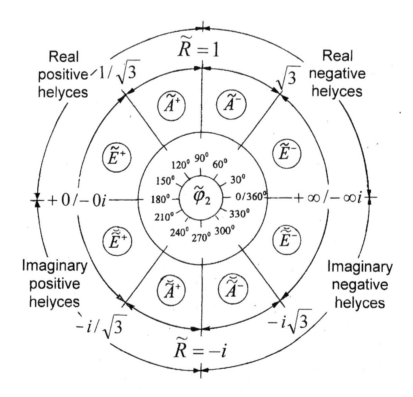

Figure 15.1.2. Main groups and subgroups of helyces as a function of helyx reality $\widetilde{R}$.

Within each main group, the helyces are further divided into two **subgroups** as shown in Table 15.1.2.

Table 15.1.2. Helyces of main groups and subgroups.

Helyces of main groups	Helyces of subgroups		Helyces of main groups	Helyces of subgroups	
Real negative	$\tilde{E}^-$	$0^0 < \tilde{\varphi}_2 < 60^0$	Imaginary positive	$\tilde{\tilde{E}}^+$	$180^0 < \tilde{\varphi}_2 < 240^0$
	$\tilde{A}^-$	$60^0 < \tilde{\varphi}_2 < 90^0$		$\tilde{\tilde{A}}^+$	$240^0 < \tilde{\varphi}_2 < 270^0$
Real positive	$\tilde{A}^+$	$90^0 < \tilde{\varphi}_2 < 120^0$	Imaginary negative	$\tilde{\tilde{A}}^-$	$270^0 < \tilde{\varphi}_2 < 300^0$
	$\tilde{E}^+$	$120^0 < \tilde{\varphi}_2 < 180^0$		$\tilde{\tilde{E}}^-$	$300^0 < \tilde{\varphi}_2 < 360^0$

15.2 Vorticities & Realities of Adjacent Helyces

Main groups of helyces - The vorticities $\tilde{V}$ and the realities $\tilde{R}$ of helyces of adjacent main groups are symmetrically related as shown in Table 15.2.1.

Table 15.2.1. Symmetrical relationships between the vorticities $\tilde{V}$ and the realities $\tilde{R}$ of polarized helyces of adjacent main groups.

Helyces of main groups		Eqs. (a)	Eqs. (b)
Reality-polarized negative helyces	$\tilde{\tilde{E}}^- \ \& \ \tilde{E}^-$ Eq. (15.2-1)	$\tilde{\tilde{V}}_E^- = 1/\tilde{V}_E^-$	$\tilde{\tilde{R}}_E^- = \pm i\tilde{R}_E^-$
Reality-polarized positive helyces	$\tilde{\tilde{E}}^+ \ \& \ \tilde{E}^+$ Eq. (15.2-2)	$\tilde{\tilde{V}}_E^+ = 1/\tilde{V}_E^+$	$\tilde{\tilde{R}}_E^+ = \pm i\tilde{R}_E^+$
Vorticity-polarized real helyces	$\tilde{\tilde{E}}^- \ \& \ \tilde{E}^-$ Eq. (15.2-3)	$\tilde{V}_A^+ = -\tilde{V}_A^-$	$\tilde{R}_A^+ = 1/\tilde{R}_A^-$
Vorticity-polarized Imaginary helyces	$\tilde{\tilde{A}}^+ \ \& \ \tilde{\tilde{A}}^-$ Eq. (15.2-4)	$\tilde{\tilde{V}}_A^+ = -\tilde{\tilde{V}}_A^-$	$\tilde{\tilde{R}}_A^+ = 1/\tilde{\tilde{R}}_A^-$

Subgroups of helyces - Table 15.2.2 summarizes the relationships between vorticities of helyces of subgroups.

Table 15.2.2. Relationships between vorticities of helyces of subgroups.

Helyces of subgroups		Equations	
Real negative helyces	$\widetilde{A}^- \ \& \ \widetilde{E}^-$	$\widetilde{V}_A^- + \widetilde{V}_E^- = -1$	(15.2-5)
Real positive helyces	$\widetilde{A}^+ \ \& \ \widetilde{E}^+$	$\widetilde{V}_A^+ + \widetilde{V}_E^+ = +1$	(15.2-6)
Imaginary positive helyces	$\widetilde{\widetilde{A}}^+ \ \& \ \widetilde{\widetilde{E}}^+$	$\dfrac{1}{\widetilde{\widetilde{V}}_A^+} + \dfrac{1}{\widetilde{\widetilde{V}}_E^+} = +1$	(15.2-7)
Imaginary negative helyces	$\widetilde{\widetilde{A}}^- \ \& \ \widetilde{\widetilde{E}}^-$	$\dfrac{1}{\widetilde{\widetilde{V}}_A^-} + \dfrac{1}{\widetilde{\widetilde{V}}_E^-} = -1$	(15.2-8)

MAIN SUMMARY

Based on their reality and vorticity, the helyces are divided into four main groups:

- *Real negative helyces*
- *Real positive helyces*
- *Imaginary positive helyces*
- *Imaginary negative helyces.*

Each main group of helyces is divided into two subgroups.

There are symmetrical relationships between parameters of helyces that belong to different groups and subgroups.

<u>Notes:</u>

16. INVERSION STATES OF HELYCES

CONTENTS

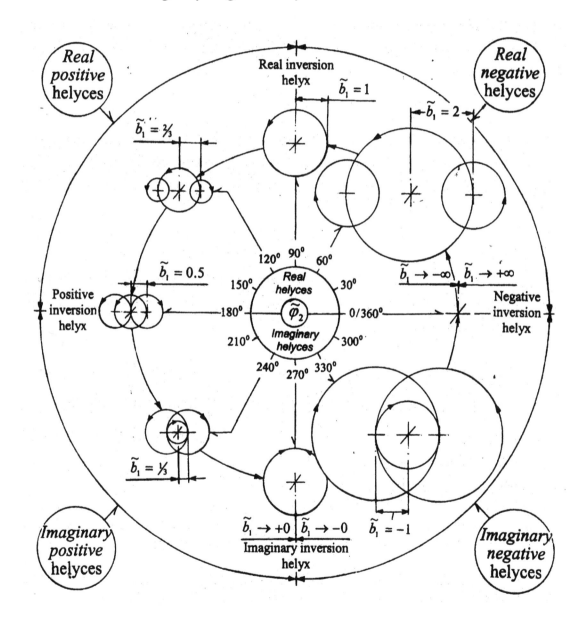

Figure 16.1.1. Metamorphoses of cross-sections of the helyx level 1 and 2 as the apex angle $\widetilde{\varphi}_2$ of the helyx level 2 increases from 0^0 to 360^0.

16.1 Inversion of Helyces

Figure 16.1.1 shows metamorphoses of cross-sections of the helyx levels 1 and 2 as the apex angle $\widetilde{\varphi}_2$ of the helyx level 2 increases from 0^0 to 360^0.

Four kinds of *inversion helyces* are located at the boundaries of the circular diagram between the four main groups of helyces.

Negative inversion helyx $(\widetilde{\varphi}_2 \rightarrow +0^0/360^0)$ - At this point, $\widetilde{b}_1 \rightarrow \pm\infty$, $\widetilde{b}_2 \rightarrow \pm\infty$ and the helyx level 1 and 2 and their wavelengths become inverted. Consequently, as $\widetilde{\varphi}_2$ crosses the borderline at $0^0/360^0$ the helyx vorticity $\widetilde{V}$ remains negative, while the helyx reality $\widetilde{R}$ inverts from imaginary to real. The helyx appears as two parallel lines separated by the distance equal to the diameter real inversion string.

Real inversion helyx $(\widetilde{\varphi}_2 \rightarrow 90^0)$ – At this point, $\widetilde{b}_1 \rightarrow +1$, $\widetilde{b}_2 \rightarrow \pm0$ and the helyx trailing string becomes inverted. Consequently, as $\widetilde{\varphi}_2$ crosses the borderline at 90^0, the helyx reality $\widetilde{R}$ remains real while its vorticity $\widetilde{V}$ inverts from negative to positive. It appears as a circle with the relative radius $\widetilde{b}_1 \rightarrow +1$.

Positive inversion helyx $(\widetilde{\varphi}_2 \rightarrow 180^0)$ – At this point, $\widetilde{b}_1 \rightarrow +\frac{1}{2}$, $\widetilde{b}_2 \rightarrow -\frac{1}{2}$ and the helyx wavelengths become inverted. Consequently, as $\widetilde{\varphi}_2$ crosses the borderline at 180^0, the helyx vorticity $\widetilde{V}$ remains positive while its reality $\widetilde{R}$ inverts from real to imaginary. At that point, the inner parts of the helyx windings are touching one another.

Imaginary inversion helyx $(\widetilde{\varphi}_2 \rightarrow 270^0)$ – At this point, $\widetilde{b}_1 \rightarrow \pm0$, $\widetilde{b}_2 \rightarrow -1$ and the helyx leading string becomes inverted. Consequently, as $\widetilde{\varphi}_2$ crosses the borderline at 270^0, the helyx reality $\widetilde{R}$ remains imaginary while its vorticity $\widetilde{V}$ inverts from positive to negative. The helyx appears as a circle with the relative radius approaching -1. The circle is located at the plane perpendicular to the plane of the real inversion string.

Figure 16.1.2 shows extreme relative parameters of inversion helyces.

Located between the inversion helyces on the circular diagram of Figure 16.1.1.1 are the helyces that belong to their four main groups. Shown below are the transformations of cross-sections of helyces within each main group.

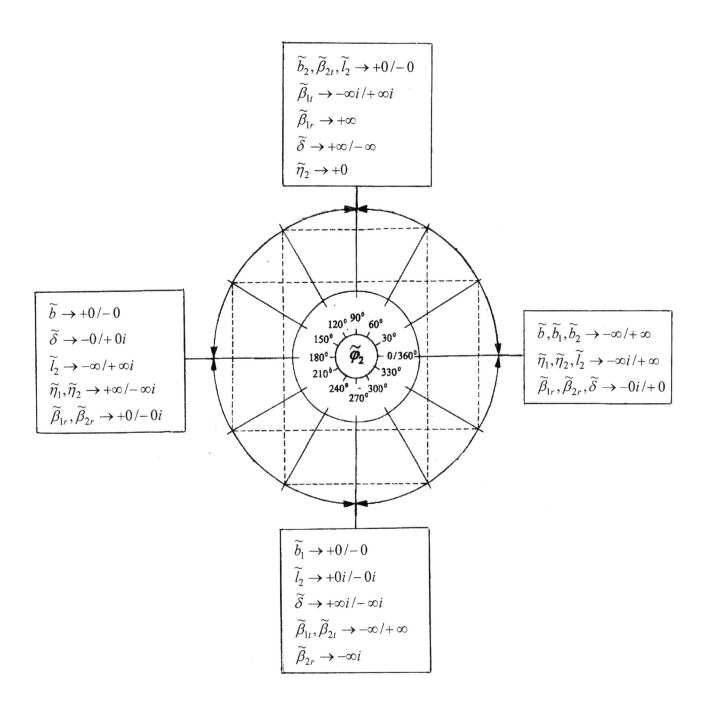

Fig. 16.1.2. Extreme relative parameters of inversion helyces.

16.2 Transformations of Real Negative Helyces

Figure 16.2 shows cross-sections of real negative helyces. They belong to the top right quadrant of the circular diagram shown in Figure 16.1.1. The trailing strings of these helyces are wound counter-clockwise outside of real inversion string. As $\widetilde{\varphi}_2$ increases, $\widetilde{b}_1$ and $\widetilde{b}_2$ decrease.

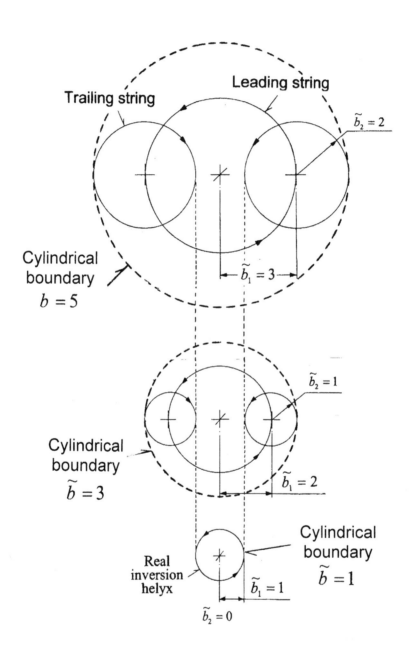

Figure 16.2. Metamorphoses of cross-sections of real negative helyces.

16.3 Transformations of Real Positive Helyces

Figure 16.3 shows cross-sections of real positive helyces. They belong to the top left quadrant of the circular diagram shown in Figure 16.1.1. Within this range, the trailing string is inverted, so that its windings are now wound clockwise inside the real inversion helyx. As $\widetilde{\varphi}_2$ increases, $\widetilde{b}_1$ decrease, but negative values of $\widetilde{b}_2$ increase. Consequently, the helyx appears as an inverted helical spiral.

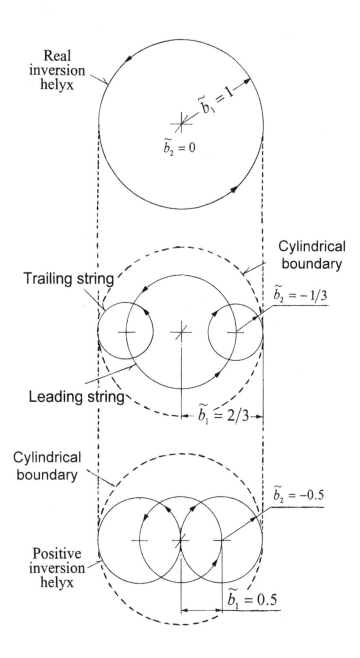

Figure 16.3. Metamorphoses of cross-sections of real positive helyces.

16.4 Transformations of Imaginary Positive Helyces

Figure 16.4 shows cross-sections of imaginary positive helyces. They belong to the bottom left quadrant of the circular diagram shown in Figure 16.1.1. As $\widetilde{\varphi}_2$ increases, $\widetilde{b}_1$ decreases, but the negative values of $\widetilde{b}_2$ increase. Within this range, the trailing string is still inverted and its windings are wound inside the imaginary inversion helyx. The opposite parts of windings of the trailing string intersect with one another.

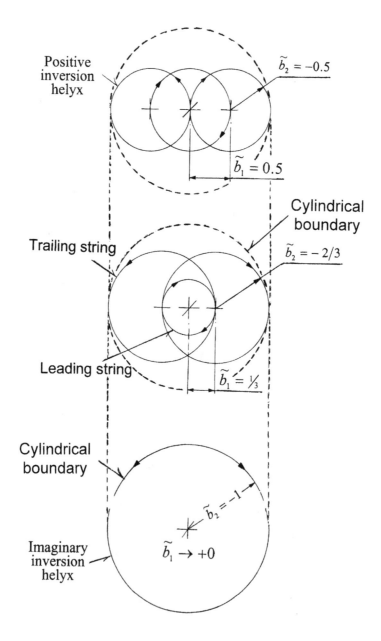

Figure 16.4. Metamorphoses of cross-sections of imaginary positive helyces.

16.5 Transformations of Imaginary Negative Helyces

Figure 16.5 shows cross-sections of imaginary negative helyces. They belong to the bottom right quadrant of circular diagram shown in Figure 16.1.1. Here the leading string becomes inverted. As $\widetilde{\varphi}_2$ increases, the negative values of $\widetilde{b}_1$ increase, the negative values of $\widetilde{b}_2$ also increase. Within this range the trailing string propagates outward and its windings are located outside of imaginary inversion helyx.

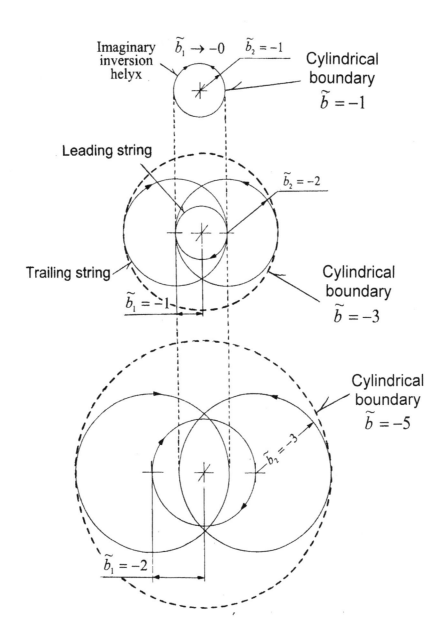

Figure 16.5. Metamorphoses of cross-sections of imaginary negative helyx.

<u>Notes</u>

17. LAMBDA RADIATION PARTICLES

CONTENTS

17.1 Formation of Radiation Particles

Radiation particles are formed when the quantum states m, n and q of their paternal trons and toryces are reduced to the lower levels. In the symbols of emitted radiation particles and their constituent helyces, the right superscripts indicate respectively the quantum states m, n and q of their parental trons and toryces prior to and after their emission of radiation particles. Preceding the right top superscripts are the signs of the radiation particle and helyx vorticities. The absence of any sign indicates that the total vorticity of the radiation particle is equal to zero. The sign preceding the radiation particle name indicates the sign of the parental particle goldicity g as shown below.

$$Electon \quad \widetilde{e}_{1,1,0}^{-1,2,0}$$

Helyx of a real electon with negative parental particle goldicity g $\quad -\widetilde{E}_{1,1,0}^{-1,2,0}$

The frequencies of emitted helyces are defined based on the ***spacetime intensity conservation law*** shown in Exhibit 17.1.

Exhibit 17.1. The spacetime intensity conservation law.

The real helyx spacetime intensity $\widetilde{I}$ is proportional to the relative frequency of real helyx trailing string $\widetilde{\delta}_2$. It is equal to a difference between the real toryx spacetime intensities I_k and I_j of its parental toryx in the higher and lower quantum states k and j:

$$\widetilde{I} = I_k - I_j = 4\alpha_s^{-1}\widetilde{\delta}_2 \qquad (17.1\text{-}1a)$$

Similarly, the imaginary helyx spacetime intensity $\breve{\widetilde{I}}$ is equal to:

$$\breve{\widetilde{I}} = \breve{I}_k - \breve{I}_j = 4\alpha_s^{-1}\breve{\widetilde{\delta}}_2 \qquad (17.1\text{-}1b)$$

Consequently, we obtain from Eqs. (17.1-1a) and (17.1-1b) that the relative frequency of real and imaginary helyx trailing strings $\widetilde{\delta}_2$ and $\widetilde{\widetilde{\delta}}_2$ are equal to:

$$\widetilde{\delta}_2 = \frac{Q_{qk}g_k - Q_{qj}g_j}{4\alpha_s^{-1}} \tag{17.1-2a}$$

$$\widetilde{\widetilde{\delta}}_2 = \frac{Q_{qk}\breve{g}_k - Q_{qj}\breve{g}_j}{4\alpha_s^{-1}} \tag{17.1-2b}$$

The energy $\widetilde{E}$ of emitted radiation particles is equal to:

$$\widetilde{E} = \widetilde{f}\,|h| \tag{17.1-3}$$

Where h = Planck constant.

Elementary radiation particles emitted by the excited trons are called *tons* and *trinos* as shown in Table 17.1. Both of them are formed when their respective polarized parental trons are transferred from higher to lower quantum states. Similar to the trons, there are two kinds of tons and trinos, the *charge-polarized tons* and *trinos* and the *reality-polarized tons and trinos*.

Table 17.1. Types of elementary radiation particles.

Parental matter trons		Elementary radiation tons		Elementary radiation trinos	
Combined	*Constituent*	*Combined*	*Constituent*	*Combined*	*Constituent*
ce-tron	Electron	ce-ton	Electon	ce-trino	Electrino
	Positron		Positon		Positrino
ca-tron	Ethertron	ca-ton	Etherton	ca-trino	Ethertrino
	Singulatrons		Singulatons		Singulatrino

The formation of two kinds of combined radiation particles are considered below: the combined ce-tons and the combined ca-tons.

Combined ce-tons - The combined ce-tons $\widetilde{ce}_{m,n,q}^{\,m,n,q}$ are emitted by the combined parental ce-trons $\widetilde{ce}_{m,n,q}^{\,0}$. The process of creation of the emitted combined ce-ton involves the following steps (Fig. 17.1.1):

- The electron $e_{m,n,q}^{-1}$ of the combined parental ce-tron $\widetilde{ce}_{m,n,q}^{0}$ emits the real and imaginary negative helyces $\widetilde{E}_{m,n,q}^{-m,n,q}$ and $\widetilde{\widetilde{E}}_{m,n,q}^{-m,n,q}$.

- The unification of the real and imaginary negative helyces $\widetilde{E}_{m,n,q}^{-m,n,q}$ and $\widetilde{\widetilde{E}}_{m,n,q}^{-m,n,q}$ produces the ***electon*** $\widetilde{e}_{m,n,q}^{-m,n,q}$.

- The positron $e_{m,n,q}^{+1}$ of the combined parental ce-tron $\widetilde{ce}_{m,n,q}^{0}$ emits the real and imaginary positive helyces $\widetilde{E}_{m,n,q}^{+m,n,q}$ and $\widetilde{\widetilde{E}}_{m,n,q}^{+m,n,q}$.

- The unification of the real and imaginary positive helyces $\widetilde{E}_{m,n,q}^{+m,n,q}$ and $\widetilde{\widetilde{E}}_{m,n,q}^{+m,n,q}$ produces the ***positon*** $\widetilde{e}_{m,n,q}^{+m,n,q}$.

- The combined ce-ton $\widetilde{ce}_{m,n,q}^{m,n,q}$ is produced by the unification of the electon $\widetilde{e}_{m,n,q}^{-m,n,q}$ with the positon $\widetilde{e}_{m,n,q}^{+m,n,q}$ both propagating simultaneously along the same path.

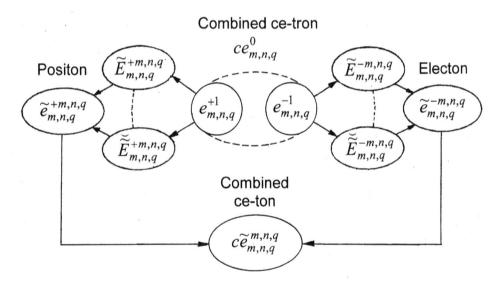

Fig. 17.1.1. Formation of the emitted combined ce-ton $\widetilde{ce}_{m,n,q}^{m,n,q}$ from the parental combined ce-tron $ce_{m,n,q}^{0}$.

Combined ca-tons – The combined ca-tons $\widetilde{ca}_{m,n,q}^{m,n,q}$ are emitted by the combined parental ca-trons $\widetilde{ca}_{m,n,q}^{0}$. The process of creation of the emitted combined ca-ton involves the following steps (see Fig. 17.1.2):

- The singulatron $\tilde{a}^0_{m,n,q}$ of the combined ca-tron $c\tilde{a}^0_{m,n,q}$ emits the imaginary positive and negative helyces $\tilde{\tilde{A}}^{+m,n,q}_{m,n,q}$ and $\tilde{\tilde{A}}^{-m,n,q}_{m,n,q}$.

- The unification of the imaginary positive and negative helyces $\tilde{\tilde{A}}^{+m,n,q}_{m,n,q}$ and $\tilde{\tilde{A}}^{-m,n,q}_{m,n,q}$ produces the ***singulaton*** $\tilde{\tilde{a}}^{m,n,q}_{m,n,q}$.

- N excited ethertrons $\tilde{a}^0_{m,n,q}$ of the combined ca-tron $c\tilde{a}^0_{m,n,q}$ emit N positive and negative helyces $\tilde{A}^{+m,n,q}_{m,n,q}$ and $\tilde{A}^{-m,n,q}_{m,n,q}$.

- The unification of N real positive and negative helyces $\tilde{A}^{+m,n,q}_{m,n,q}$ and $\tilde{A}^{-m,n,q}_{m,n,q}$ produces N ***ethertons*** $\tilde{a}^{m,n,q}_{m,n,q}$.

- The combined ca-ton $c\tilde{a}^{m,n,q}_{m,n,q}$ is produced by the unification of one singulaton $\tilde{\tilde{a}}^{m,n,q}_{m,n,q}$ with N ethertons $\tilde{a}^{m,n,q}_{m,n,q}$ propagating simultaneously along the same path.

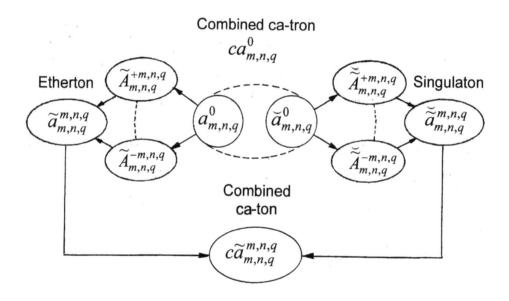

Fig. 17.1.2. Formation of the emitted combined ca-ton $c\tilde{a}^{m,n,q}_{m,n,q}$ from the combined parental ca-tron $c\tilde{a}^0_{m,n,q}$.

17.2 Lambda Electons & Positons

17.2.1 Parental lambda electrons– see Table 17.2.1.

Table 17.2.1. Compositions and properties of parental lambda electrons of the spacetime levels L with the quantum states (m) when $Q_q = 1$.

L (m)	Parental electrons	Toryces of parental lambda electrons					
		Symbols	b_1	$\beta_1 = \beta_{2t}$	β_{2r}	δ_2	g
$L0$ (1)	$e_{1,1,0}^{-1}$	$E_{1,1,0}^-$	274.00	0.08535778	0.99635036	3.6496×10^{-3}	-7.3126×10^{-3}
		$\breve{E}_{1,1,0}^-$	-273.00	$-0.08567044i$	1.00366300	-3.6630×10^{-3}	7.3126×10^{-3}
	$e_{1,2,0}^{-1}$	$E_{1,2,0}^-$	548.00	0.06038464	0.99817518	1.8248×10^{-3}	-3.6530×10^{-3}
		$\breve{E}_{1,2,0}^-$	-547.00	$-0.06049504i$	1.00182815	-1.8282×10^{-3}	3.6530×10^{-3}
	$e_{1,3,0}^{-1}$	$E_{1,3,0}^-$	822.00	0.04931136	0.99878346	1.2166×10^{-3}	-2.4346×10^{-3}
		$\breve{E}_{1,3,0}^-$	-821.00	$-0.04937142i$	1.00181218	-1.2180×10^{-3}	2.4346×10^{-3}
$L1$ (2)	$e_{2,1,0}^{-1}$	$E_{2,1,0}^-$	37538.0	0.00729922	0.99997336	2.6640×10^{-5}	-5.3280×10^{-5}
		$\breve{E}_{2,1,0}^-$	-37537.0	$-0.00729942i$	1.00002664	-2.6640×10^{-5}	5.3280×10^{-5}
	$e_{2,2,0}^{-1}$	$E_{2,2,0}^-$	150152.0	0.00364963	0.99999334	6.6600×10^{-6}	-1.3320×10^{-5}
		$\breve{E}_{2,2,0}^-$	-150151.0	$-0.00364965i$	1.00000666	-6.6600×10^{-6}	1.3320×10^{-5}
	$e_{2,3,0}^{-1}$	$E_{2,3,0}^-$	337842.0	0.00243309	0.99999970	2.9600×10^{-6}	-5.9120×10^{-6}
		$\breve{E}_{2,3,0}^-$	-337841.0	$-0.00243310i$	1.00000296	-2.9600×10^{-6}	5.9120×10^{-6}
$L2$ (3)	$e_{3,1,0}^{-1}$	$E_{3,1,0}^-$	5142706.0	0.00062362	0.99999981	1.9445×10^{-7}	-3.8890×10^{-7}
		$\breve{E}_{3,1,0}^-$	-5142705.0	$-0.00062362i$	1.00000019	-1.9445×10^{-7}	3.8890×10^{-7}
	$e_{3,2,0}^{-1}$	$E_{3,2,0}^-$	41141648.0	0.00022048	0.99999998	2.4306×10^{-8}	-4.8612×10^{-8}
		$\breve{E}_{3,2,0}^-$	-41141647.0	$-0.00022048i$	1.00000002	-2.4306×10^{-8}	4.8612×10^{-8}
	$e_{3,3,0}^{-1}$	$E_{3,3,0}^-$	138853062.0	0.00012002	0.99999993	7.2018×10^{-9}	-1.4404×10^{-8}
		$\breve{E}_{3,3,0}^-$	-138853061.0	$-0.00012002i$	1.00000007	-7.2018×10^{-9}	1.4404×10^{-8}

17.2.2 Emitted lambda electons– See Table 17.2.2.

Table 17.2.2. Compositions and properties of lambda electons emitted in the spacetime levels L with the quantum states (m).

L (m)	Emitted electons	\multicolumn{6}{c}{Helyces of emitted lambda electons}
		Symbols	g	$\widetilde{b}_1$	$\widetilde{\beta}_{1t}$	$\widetilde{f}$, Hz	$\widetilde{E}^-$, MeV
L0 (1)	$\widetilde{e}_{1,1,0}^{-1,2,0}$	$-\widetilde{E}_{1,1,0}^{-1,2,0}$	-7.3126×10^{-3}	3.5539×10^3	0.99943700	2.26×10^{17}	1.87×10^{-3}
		$+\widetilde{E}_{1,1,0}^{-1,2,0}$	7.3126×10^{-3}	-3.5539×10^3	1.00056268	-2.26×10^{17}	
	$\widetilde{e}_{1,2,0}^{-1,3,0}$	$-\widetilde{E}_{1,2,0}^{-1,3,0}$	-3.6530×10^{-3}	7.3979×10^3	0.99972960	7.53×10^{16}	6.22×10^{-4}
		$+\widetilde{E}_{1,2,0}^{-1,3,0}$	3.6530×10^{-3}	-7.3979×10^3	1.00027033	-7.53×10^{16}	
L1 (2)	$\widetilde{e}_{2,1,0}^{-2,2,0}$	$-\widetilde{E}_{2,1,0}^{-2,2,0}$	-2.4346×10^{-3}	7.2197×10^4	0.99997230	2.47×10^{15}	2.04×10^{-5}
		$+\widetilde{E}_{2,1,0}^{-2,2,0}$	2.4346×10^{-3}	-7.2197×10^4	1.00002770	-2.47×10^{15}	
	$\widetilde{e}_{2,2,0}^{-2,3,0}$	$-\widetilde{E}_{2,2,0}^{-2,3,0}$	-5.3280×10^{-5}	2.2222×10^5	0.99999100	4.57×10^{14}	3.78×10^{-6}
		$+\widetilde{E}_{2,2,0}^{-2,3,0}$	5.3280×10^{-5}	-2.2222×10^5	1.00000900	-4.57×10^{14}	
L2 (3)	$\widetilde{e}_{3,1,0}^{-3,2,0}$	$-\widetilde{E}_{3,1,0}^{-3,2,0}$	-1.3320×10^{-5}	1.7313×10^6	0.99999884	2.10×10^{13}	1.74×10^{-7}
		$+\widetilde{E}_{3,1,0}^{-3,2,0}$	1.3320×10^{-5}	-1.7313×10^6	1.00000116	-2.10×10^{13}	
	$\widetilde{e}_{3,2,0}^{-3,3,0}$	$-\widetilde{E}_{3,2,0}^{-3,3,0}$	-5.9120×10^{-6}	8.0078×10^6	0.99999975	2.11×10^{12}	1.75×10^{-8}
		$+\widetilde{E}_{3,2,0}^{-3,3,0}$	5.9120×10^{-6}	-8.0078×10^6	1.00000025	-2.11×10^{12}	

17.2.3 Parental lambda positrons – See Table 17.2.3.

Table 17.2.3. Compositions and properties of parental lambda positrons of the spacetime levels L with the quantum states (m) when $Q_q = 1$.

L (m)	Parental positrons	Toryces of parental lambda positrons					
		Symbols	b_1	$\beta_1 = \beta_{2t}$	β_{2r}	δ_2	g
L0 (1)	$e_{1,1,0}^{+1}$	$E_{1,1,0}^{+}$	0.50091408	0.08535778	-0.99635036	1.99635036	7.3126×10^{-3}
		$\breve{E}_{1,1,0}^{+}$	0.49908592	$0.08567044i$	-1.00366300	2.00366300	-7.3126×10^{-3}
	$e_{1,2,0}^{+1}$	$E_{1,2,0}^{+}$	0.50045662	0.06038464	-0.99817518	1.99817518	3.6530×10^{-3}
		$\breve{E}_{1,2,0}^{+}$	0.49954338	$0.06049504i$	-1.00182815	2.00182815	-3.6530×10^{-3}
	$e_{1,3,0}^{+1}$	$E_{1,3,0}^{+}$	0.50030432	0.04931136	-0.99878346	1.99878345	2.4346×10^{-3}
		$\breve{E}_{1,3,0}^{+}$	0.49969568	$0.04937142i$	-1.00181218	2.00121803	-2.4346×10^{-3}
L1 (2)	$e_{2,1,0}^{+1}$	$E_{2,1,0}^{+}$	0.50000666	0.00729922	-0.99997336	1.99997336	5.3280×10^{-5}
		$\breve{E}_{2,1,0}^{+}$	0.49999334	$0.00729942i$	-1.00002664	2.00002664	-5.3280×10^{-5}
	$e_{2,2,0}^{+1}$	$E_{2,2,0}^{+}$	0.50000166	0.00364963	-0.99999334	1.99999334	1.3320×10^{-5}
		$\breve{E}_{2,2,0}^{+}$	0.49999834	$0.00364965i$	-1.00000666	2.00000666	-1.3320×10^{-5}
	$e_{2,3,0}^{+1}$	$E_{2,3,0}^{+}$	0.50000074	0.00243309	-0.99999704	1.99999704	5.9120×10^{-6}
		$\breve{E}_{2,3,0}^{+}$	0.49999926	$0.00243310i$	-1.00000296	2.00000296	-5.9120×10^{-6}
L2 (3)	$e_{3,1,0}^{+1}$	$E_{3,1,0}^{+}$	0.5000000486	0.00062362	-0.99999981	1.999999805	3.8890×10^{-7}
		$\breve{E}_{3,1,0}^{+}$	0.4999999514	$0.00062362i$	-1.00000019	2.000000195	-3.8890×10^{-7}
	$e_{3,2,0}^{+1}$	$E_{3,2,0}^{+}$	0.5000000061	0.00022048	-0.99999998	1.999999976	4.8612×10^{-8}
		$\breve{E}_{3,2,0}^{+}$	0.4999999939	$0.00022048i$	-1.00000002	2.000000024	-4.8612×10^{-8}
	$e_{3,3,0}^{+1}$	$E_{3,3,0}^{+}$	0.5000000018	0.00012002	-0.99999999	1.999999999	1.4404×10^{-8}
		$\breve{E}_{3,3,0}^{+}$	0.4999999982	$0.00012002i$	-1.00000001	2.000000001	-1.4404×10^{-8}

17.2.4 Emitted lambda positons – See Table 17.2.4.

Table 17.2.4. Compositions and properties of lambda positons emitted in the spacetime levels L with the quantum states (m).

L (m)	Emitted positons	Symbols	g	$\tilde{b}_1$	$\tilde{\beta}_{1t}$	$\tilde{f}$, Hz	$\tilde{E}^+$, MeV
				Helyces of emitted lambda positons			
$L0$ (1)	$\tilde{e}_{1,1,0}^{+1,2,0}$	$+\tilde{E}_{1,1,0}^{+1,2,0}$	7.3126×10^{-3}	-3.553×10^{3}	1.00056268	-2.26×10^{17}	1.87×10^{-3}
		$-\tilde{E}_{1,1,0}^{+1,2,0}$	-7.3126×10^{-3}	3.554×10^{3}	0.99943700	2.26×10^{17}	
	$\tilde{e}_{1,2,0}^{+1,3,0}$	$+\tilde{E}_{1,2,0}^{+1,3,0}$	3.6530×10^{-3}	-7.397×10^{3}	1.00027033	-7.53×10^{16}	6.22×10^{-4}
		$-\tilde{E}_{1,2,0}^{+1,3,0}$	-3.6530×10^{-3}	7.398×10^{3}	0.99972960	7.53×10^{16}	
$L1$ (2)	$\tilde{e}_{2,1,0}^{+2,2,0}$	$+\tilde{E}_{2,1,0}^{+2,2,0}$	2.4346×10^{-3}	-7.220×10^{4}	1.00002770	-2.47×10^{15}	2.04×10^{-5}
		$-\tilde{E}_{2,1,0}^{+2,2,0}$	-2.4346×10^{-3}	7.220×10^{4}	0.99997230	2.47×10^{15}	
	$\tilde{e}_{2,2,0}^{+2,3,0}$	$+\tilde{E}_{2,2,0}^{+2,3,0}$	5.3280×10^{-5}	-2.222×10^{5}	1.00000900	-4.57×10^{14}	3.78×10^{-6}
		$-\tilde{E}_{2,2,0}^{-2,3,0}$	-5.3280×10^{-5}	2.222×10^{5}	0.99999100	4.57×10^{14}	
$L2$ (3)	$\tilde{e}_{3,1,0}^{+3,2,0}$	$+\tilde{E}_{3,1,0}^{+3,2,0}$	1.3320×10^{-5}	-1.731×10^{6}	1.00000116	-2.10×10^{13}	1.74×10^{-7}
		$-\tilde{E}_{3,1,0}^{+3,2,0}$	-1.3320×10^{-5}	1.731×10^{6}	0.99999884	2.10×10^{13}	
	$\tilde{e}_{3,2,0}^{+3,3,0}$	$+\tilde{E}_{3,2,0}^{+3,3,0}$	5.9120×10^{-6}	-8.001×10^{6}	1.00000025	-2.11×10^{12}	1.75×10^{-8}
		$-\tilde{E}_{3,2,0}^{+3,3,0}$	-5.9120×10^{-6}	8.001×10^{6}	0.99999975	2.11×10^{12}	

17.2.5 Analysis of calculated data for emitted lambda electons & positons - It follows from Tables 17.2.2. 17.2.4 and Figs. 17.2.1 – 17.2.3:

- Real and imaginary helyces making up both lambda electons and positons are emitted in the same direction (Fig. 17.2.1).
- The translational velocities of leading strings $\tilde{\beta}_{1t}$ of emitted real helyces making up excited electons and positons are slightly less than velocity of light, while these velocities of emitted imaginary helyces are slightly greater than velocity of light.

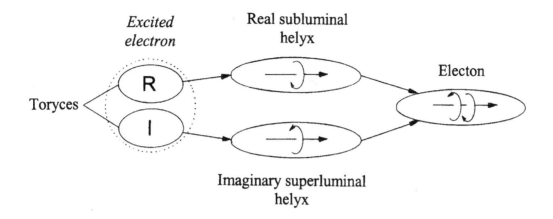

Fig. 17.2.1. Formation of emitted electons.

The following changes of parameters of trailing strings of constituent helyces of lambda electons and positons occur as the spacetime level L increases:

- The relative radii $\widetilde{b}_1$ increase
- The frequencies $\widetilde{f}_2$ decrease
- The differences between the translational velocities of trailing strings $\widetilde{\beta}_{2t}$ of real and imaginary helyces decrease.

When the linear excitation quantum states n of parental electrons and positrons are reduced, the calculated frequencies of emitted electons and positons $\widetilde{f}_2$ of various spacetime levels L are within the following frequency ranges:

- For the spacetime level *L0* ($m = 1$), the calculated frequencies are within the frequency range of cosmic **X-ray background (CXB) radiations**.
- For the spacetime level *L1* ($m = 2$), the calculated frequencies are within the frequency ranges of **infrared, visible** and **ultraviolet radiations**.
- For the spacetime level *L2* ($m = 3$), the calculated frequencies are within the frequency ranges of infrared and microwave frequency ranges known in astronomy as **cosmic microwave background (CMB) radiation**.

Experimental data for the frequencies $\widetilde{f}_2$ of some spectra lines of a hydrogen atom are very accurately described by the Rydberg's Eq. (7.2-1), Figs. 17.2.2, 17.2.3 and Table 7.2.4).

$$\widetilde{f}_2 = R_\infty c \left(\frac{1}{n_j^2} - \frac{1}{n_k^2} \right) \tag{17.2-1}$$

where R_∞ is the Rydberg constant.

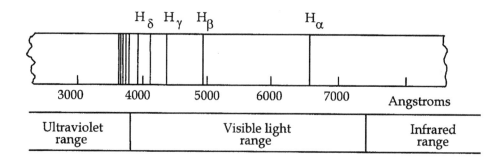

Figure 17.2.2. The line spectrum for atomic hydrogen.
Adapted from G. Gamow (1985).

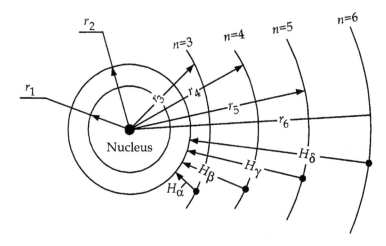

Figure 17.2.3. Electron orbits and spectra lines of a hydrogen atom.

Table 17.2.4. Comparison of the calculated frequencies $\widetilde{f}_2$ of emitted lambda electon-positons with frequencies of spectra lines for hydrogen atom calculated from the Rydberg's equation.

Spectra lines of hydrogen	Quantum states $(k-j)$	Excited ce-tons	Frequencies $\widetilde{f}_2$, Hz		Calculated/ Rydbers's
			Calculated	Rydberg's	
H_α	$(3-2)$	$c\widetilde{e}_{2,2,0}^{2,3,0}$	4.571648×10^{14}	4.569225×10^{14}	1.000530
H_β	$(4-2)$	$c\widetilde{e}_{2,2,0}^{2,4,0}$	6.171721×10^{14}	6.168454×10^{14}	1.000530
H_γ	$(5-2)$	$c\widetilde{e}_{2,2,0}^{2,5,0}$	6.912325×10^{14}	6.908668×10^{14}	1.000529
H_δ	$(6-2)$	$c\widetilde{e}_{2,2,0}^{2,6,0}$	7.314629×10^{14}	7.310760×10^{14}	1.000529

17.3 Lambda Singulatons & Ethertons

17.3.1 Parental lambda singulatrons – See Table 17.3.1.

Table 17.3.1. Compositions and properties of parental lambda singulatrons of the spacetime levels L with the quantum states (m).

L (m)	Singula-trons	Symbols	b_1	$\beta_1 = \beta_{2t}$	β_{2r}	δ_2	g
				Toryces of parental lambda singulatrons			
$L0$ (0)	$\breve{a}_{0,1,0}^{0}$	$\breve{A}_{0,1,0}^{+}$	0.33333333	1.73205081i	-2.00000000	3.00000000	-1.50000000
		$\breve{A}_{0,1,0}^{-}$	-1.00000000	-1.73205081i	2.00000000	-1.00000000	1.50000000
$L1$ (1)	$\breve{a}_{1,1,0}^{0}$	$\breve{A}_{1,1,0}^{+}$	0.00363636	273.998175i	-274.000000	275.000000	-273.996350
		$\breve{A}_{1,1,0}^{-}$	-0.00366300	-273.998175i	274.000000	-273.000000	273.996350
	$\breve{a}_{1,2,0}^{0}$	$\breve{A}_{1,2,0}^{+}$	0.00182149	547.999088i	-548.000000	549.000000	-574.998175
		$\breve{A}_{1,2,0}^{-}$	-0.00182815	-547.999088i	548.000000	-547.000000	574.998175
	$\breve{a}_{1,3,0}^{0}$	$\breve{A}_{1,3,0}^{+}$	0.00121507	821.999392i	-822.000000	823.000000	-821.998784
		$\breve{A}_{1,3,0}^{-}$	-0.00121803	-821.999392i	822.000000	-821.000000	821.998784
	$\breve{a}_{1,4,0}^{0}$	$\breve{A}_{1,4,0}^{+}$	0.00091324	1095.99954i	-1096.00000	1095.00000	-1095.99909
		$\breve{A}_{1,4,0}^{-}$	-0.00091324	-1095.99954i	1096.00000	-1095.00000	1095.99909
$L2$ (2)	$\breve{a}_{2,1,0}^{0}$	$\breve{A}_{2,1,0}^{+}$	0.00002664	37538.0000i	-37538.0000	37539.0000	-37538.000
		$\breve{A}_{2,1,0}^{-}$	-0.00002664	-37538.0000i	37538.0000	-37537.0000	37538.000
	$\breve{a}_{2,2,0}^{0}$	$\breve{A}_{2,2,0}^{+}$	0.00000666	152152.000i	-152152.000	152153.000	-152152.000
		$\breve{A}_{2,2,0}^{-}$	-0.00000666	-152152.000i	152152.000	-152151.000	152152.000
	$\breve{a}_{2,3,0}^{0}$	$\breve{A}_{2,3,0}^{+}$	0.00000296	337842.000i	-337842.000	337843.000	-337842.000
		$\breve{A}_{2,3,0}^{-}$	-0.00000296	-337842.000i	337842.000	-337841.000	337842.000
	$\breve{a}_{2,4,0}^{0}$	$\breve{A}_{2,4,0}^{+}$	0.00000166	600608.000i	-822.000000	600607.000	-600608.000
		$\breve{A}_{2,4,0}^{-}$	-0.00000166	-600608.000i	822.000000	-600607.000	600608.000

17.3.2 Emitted lambda singulatons – See Table 17.3.2.

Table 17.3.2. Compositions and properties of lambda singulatons emitted in the spacetime levels L with the quantum states (m).

L (m)	Emitted singula-tons	Helyces of emitted lambda singulatons					
		Symbols	g	$\tilde{b}_1$	$\tilde{\beta}_{1t}$	$\tilde{f}$, Hz	$\tilde{E}^0$, GeV
$L0$ (0)	$\tilde{a}_{0,1,0}^{0,2,0}$	$-\tilde{A}_{0,1,0}^{0,2,0}$	-1.50000000	$\to +\infty$	1.0000000	0.0000000	0.0000000
		$+\tilde{A}_{0,1,0}^{0,2,0}$	1.50000000	$\to -\infty$	1.0000000	0.0000000	
$L1$ (1)	$\tilde{a}_{1,1,0}^{1,2,0}$	$-\tilde{A}_{1,1,0}^{1,2,0}$	-273.996350	0.50780469	-0.93543230	-1.693×10^{22}	0.1400146
		$+\tilde{A}_{1,1,0}^{1,2,0}$	273.996350	0.49219531	-1.00194828	1.693×10^{22}	
	$\tilde{a}_{1,2,0}^{1,3,0}$	$-\tilde{A}_{1,2,0}^{1,3,0}$	-574.998175	0.50780462	-0.93543290	-1.693×10^{22}	0.1400140
		$+\tilde{A}_{1,2,0}^{1,3,0}$	574.998175	0.49219538	-1.00194824	1.693×10^{22}	
	$\tilde{a}_{1,3,0}^{1,4,0}$	$-\tilde{A}_{1,3,0}^{1,4,0}$	-821.998784	0.50780461	-0.93543304	-1.693×10^{22}	0.1400138
		$+\tilde{A}_{1,3,0}^{1,4,0}$	821.998784	0.49219539	-1.00194823	1.693×10^{22}	
$L2$ (2)	$\tilde{a}_{2,1,0}^{2,2,0}$	$-\tilde{A}_{2,1,0}^{2,2,0}$	-1095.99909	0.00486746	-204.443545	-6.957×10^{24}	57.545600
		$+\tilde{A}_{2,1,0}^{2,2,0}$	1095.99909	-0.00486746	206.443569	6.957×10^{24}	
	$\tilde{a}_{2,2,0}^{2,3,0}$	$-\tilde{A}_{2,2,0}^{2,3,0}$	-37538.000	0.00292046	341.408520i	-1.160×10^{25}	95.909332
		$+\tilde{A}_{2,2,0}^{2,3,0}$	37538.000	-0.00292046	-343.408528i	1.160×10^{25}	
	$\tilde{a}_{2,3,0}^{2,4,0}$	$-\tilde{A}_{2,3,0}^{2,4,0}$	-152152.000	0.00208605	478.372933i	-1.623×10^{25}	134.273066
		$-\tilde{A}_{2,3,0}^{2,4,0}$	152152.000	-0.00208605	-480.372937i	1.160×10^{25}	

17.3.3 Parental lambda ethertrons – See Table 17.3.3.

Table 17.3.3. Compositions and properties of parental lambda ethertrons of the spacetime levels L with the quantum states (m).

L (m)	Ethertrons	Toryces of parental lambda ethertrons					
		Symbols	b_1	$\beta_1 = \beta_{2t}$	β_{2r}	δ_2	g
$L0$ (0)	$a_{0,1,0}^0$	$A_{0,1,0}^-$	2.00000000	0.86602540	0.50000000	0.50000000	-1.50000000
		$A_{0,1,0}^+$	0.66666667	0.86602540	-0.50000000	1.50000000	1.50000000
$L1$ (1)	$a_{1,1,0}^0$	$A_{1,1,0}^-$	1.00366300	0.99999334	0.00364964	0.99635036	-273.996350
		$A_{1,1,0}^+$	0.99636364	0.99999334	-0.00364964	1.00364964	273.996350
	$a_{1,2,0}^0$	$A_{1,2,0}^-$	1.00182815	0.99999834	0.00182482	0.99817518	-574.998175
		$A_{1,2,0}^+$	0.99817851	0.99999834	0.00182482	1.00182482	574.998175
	$a_{1,3,0}^0$	$A_{1,3,0}^-$	1.00121803	0.99999926	0.00121655	0.99878345	-821.998783
		$A_{1,3,0}^+$	-0.99878493	0.99999926	-0.00121655	1.00121655	821.998784
	$a_{1,4,0}^0$	$A_{1,4,0}^-$	1.00091324	0.99999958	0.00091241	0.99908759	-1095.99909
		$A_{1,4,0}^+$	-0.99908842	0.99999958	0.00091241	1.00091241	1095.99909
$L2$ (2)	$a_{2,1,0}^0$	$A_{2,1,0}^-$	1.00002664	0.99999999	0.00002664	0.99997336	-37538.000
		$A_{2,1,0}^+$	0.99997336	0.99999999	-0.00002664	1.00002664	37538.000
	$a_{2,2,0}^0$	$A_{2,2,0}^-$	1.00000666	0.99999999	0.00000666	0.99999334	-152152.000
		$A_{2,2,0}^+$	0.99999334	0.99999999	-0.00000666	1.00000666	152152.000
	$a_{2,3,0}^0$	$A_{2,3,0}^-$	1.00000296	0.99999999	0.00000296	0.99999704	-337842.000
		$A_{2,3,0}^+$	0.99999704	0.99999999	-0.00000296	1.00000296	337842.000
	$a_{2,4,0}^0$	$A_{2,4,0}^-$	1.00000166	0.99999999	0.00000166	0.99999834	-600608.000
		$A_{2,4,0}^+$	0.99999834	0.99999999	0.00000166	1.00000166	600608.000

17.3.4 Emitted lambda ethertons – See Table 17.3.4.

Table 17.3.4. Compositions and properties of lambda ethertons emitted in the spacetime levels L with the quantum states (m).

L (m)	Emitted ethertons	Helyces of emitted lambda ethertons					
		Symbols	g	$\widetilde{b}_1$	$\widetilde{\beta}_{1t}$	$\widetilde{f}$, Hz	$\widetilde{E}^0$, GeV
$L0$ (0)	$\widetilde{a}_{0,1,0}^{0,2,0}$	$+\widetilde{A}_{0,1,0}^{0,2,0}$	1.50000000	$\to +\infty$	1.0000000	0.0000000	0.0000000
		$-\widetilde{A}_{0,1,0}^{0,2,0}$	-1.50000000	$\to -\infty$	1.0000000	0.0000000	
$L1$ (1)	$\widetilde{a}_{1,1,0}^{1,2,0}$	$+\widetilde{A}_{1,1,0}^{1,2,0}$	273.996350	0.50780469	-0.93543230	-1.693×10^{22}	0.1400146
		$-\widetilde{A}_{1,1,0}^{1,2,0}$	-273.996350	0.49219531	-1.00194828	1.693×10^{22}	
	$\widetilde{a}_{1,2,0}^{1,3,0}$	$+\widetilde{A}_{1,2,0}^{1,3,0}$	574.998175	0.50780462	-0.93543290	-1.693×10^{22}	0.1400140
		$-\widetilde{A}_{1,2,0}^{1,3,0}$	-574.998175	0.49219538	-1.00194824	1.693×10^{22}	
	$\widetilde{a}_{1,3,0}^{1,4,0}$	$+\widetilde{A}_{1,3,0}^{1,4,0}$	821.998784	0.50780461	-0.93543304	-1.693×10^{22}	0.1400138
		$-\widetilde{A}_{1,3,0}^{1,4,0}$	-821.998783	0.49219539	-1.00194823	1.693×10^{22}	
$L2$ (2)	$\widetilde{a}_{2,1,0}^{2,2,0}$	$+\widetilde{A}_{2,1,0}^{2,2,0}$	1095.99909	0.00486746	-204.443545	-6.957×10^{24}	57.545600
		$-\widetilde{A}_{2,1,0}^{2,2,0}$	-1095.99909	-0.00486746	206.443569	6.957×10^{24}	
	$\widetilde{a}_{2,2,0}^{2,3,0}$	$+\widetilde{A}_{2,2,0}^{2,3,0}$	37538.000	0.00292046	$341.408520i$	-1.160×10^{25}	95.909332
		$-\widetilde{A}_{2,2,0}^{-2,3,0}$	-37538.000	-0.00292046	$-343.408528i$	1.160×10^{25}	
	$\widetilde{a}_{2,3,0}^{2,4,0}$	$+\widetilde{A}_{2,3,0}^{2,4,0}$	152152.000	0.00208605	$478.372933i$	-1.623×10^{25}	134.273066
		$-\widetilde{A}_{2,3,0}^{-2,4,0}$	-152152.000	-0.00208605	$-480.372937i$	1.160×10^{25}	

17.3.5 Analysis of calculated data for emitted lambda singulatons & ethertons - It follows from Tables 17.3.1 - 17.3.4 and Fig. 17.3.1:

- Negative and positive lambda singulatons and ethertons of the spacetime level *L1* are emitted in same directions similarly to the electrons and positrons (Fig. 17.2.1). Their average translational velocities are equal to velocity of light and the frequencies are within the range of **gamma rays**.
- Negative and positive lambda singulatons and ethertons of the spacetime level *L2* and higher are emitted in opposite directions (Fig. 17.3.1). Their translational velocities exceed significantly velocity of light and the frequencies exceed 10^{24} Hz.

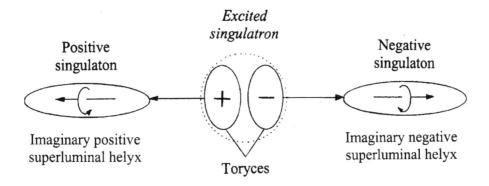

Fig. 17.3.1. Formation of emitted lambda singulatons of the spacetime level *L2* and higher.

The following changes of parameters of trailing strings of constituent helyces of lambda lambda singulatons and ethertons occur as the spacetime level *L* increases:

- The frequencies $\widetilde{f}_2$ increase
- The differences between the translational velocities of trailing strings $\widetilde{\beta}_{2t}$ of real and imaginary helyces decrease.

Combined ethertons and singulatons $c\widetilde{a}_{1,n,q}^{1,n,q}$ of the spacetime level *L1* ($m = 1$) decay according to the following sequence:

$$c\widetilde{a}_{1,3,0}^{1,4,0} \rightarrow c\widetilde{a}_{1,2,0}^{1,3,0} \rightarrow c\widetilde{a}_{1,1,0}^{1,2,0} \qquad (17.3\text{-}1)$$

$$140.0146\,MeV \quad 140.0140\,MeV \quad 140.0138\,MeV$$

Notably, the energy of the combined ethertons and singulatons $c\widetilde{a}_{1,1,0}^{1,2,0}$ of $140.0138\,MeV$ is about 3.7% greater than the measured energy $134.9766\,MeV$ of the neutral pion π^0.

Combined ethertons and singulatons $c\widetilde{a}_{m,n,q}^{m,n,q}$ of the spacetime level $L2$ ($m = 2$) decay according to the following sequence:

$$c\widetilde{a}_{2,3,0}^{2,4,0} \rightarrow c\widetilde{a}_{2,2,0}^{2,3,0} \rightarrow c\widetilde{a}_{2,1,0}^{2,2,0}$$

(17.3-2)

$$134.273 \, GeV \quad 95.909 \, GeV \quad 57.546 \, GeV$$

Notably, the energy of the combined ethertons and singulatons $c\widetilde{a}_{2,3,0}^{2,4,0}$ of $134.273 \, GeV$ is about 7.3% greater than the energy $125.09 \, GeV$ of the Higgs boson reported by CERN in 2017. The energy of the combined ethertons and singulatons $c\widetilde{a}_{2,2,0}^{2,3,0}$ of $95.909 \, GeV$ is about 5.1% greater than the energy $91.206 \, GeV$ of the Z boson reported by CERN in 2017.

MAIN SUMMARY

Real and imaginary helyces making up lambda electons and positons propagate in the same direction for all spacetime levels of our Multiverse.

Velocities of propagation of lambda ce-tons are equal to the velocity of light for all spacetime levels of our Multiverse. (Tables 17.2.2 and 17.2.4).

Real and imaginary helyces making up lambda singulatons and ethertons propagate in the same direction for the spacetime levels L0 and L1 of our Multiverse. These directions are opposite for the spacetime levels of our Multiverse greater than L1.

Velocities of propagation of lambda ca-tons are equal to the velocity of light for the spacetime levels L0 and L1 of the Multiverse. These velocities are superluminal for the spacetime levels of our Multiverse greater than L1 (Tables 17.3.2 and 17.3.4).

*The following matter particles do not emit radiation and can be qualified as a **dark matter** of the spacetime levels of our Multiverse:*

- *Lambda singulatrons of the spacetime level L0 (Tables 17.3.1 and 17.3.2)*
- *Lambda ethertrons of the spacetime level L0 (Tables 17.3.3 and 17.3.4).*

Notes:

18. HARMONIC & OSCILLATED RADIATION PARTICLES

CONTENTS

18.1 Harmonic Electons & Positons

18.1.1 Parental harmonic electrons – see Table 18.1.1.

Table 18.1.1. Compositions and properties of parental harmonic electrons of all spacetime levels L with the quantum states $m = 0$.

Parental electrons	Toryces of parental harmonic electrons					
	Symbols	b_1	$\beta_1 = \beta_{2t}$	β_{2r}	δ_2	g
$e_{H,0,0}^{-1}$	$E_{H,0,0}^{-1/2}$	2.00	0.86602540	0.50000000	0.50000000	-1.5000000
	$\breve{E}_{H,0,0}^{-2}$	-1.00	-1.73205081i	2.00000000	-1.00000000	1.5000000
$e_{H,1,0}^{-1}$	$E_{H,1,0}^{-2/3}$	3.00	0.74535599	0.66666667	0.33333333	-0.8333333
	$\breve{E}_{H,1,0}^{-3/2}$	-2.00	-1.11803399i	1.50000000	-0.50000000	0.8333333
$e_{H,2,0}^{-1}$	$E_{H,2,0}^{-3/4}$	4.00	0.66143783	0.75000000	0.25000000	-0.5833333
	$\breve{E}_{H,2,0}^{-4/3}$	-3.00	-0.88191710i	1.33333333	-0.33333333	0.5833333
$e_{H,3,0}^{-1}$	$E_{H,3,0}^{-4/5}$	5.0	0.60000000	0.80000000	0.20000000	-0.4500000
	$\breve{E}_{H,3,0}^{-5/4}$	-4.0	-0.75000000i	1.25000000	-0.25000000	0.4500000

18.1.2 Harmonic electons – see Table 18.1.2.

Table 18.1.2. Compositions and properties of harmonic electons emitted by their parental electrons in all spacetime levels L with the quantum states $m = 0$.

Parental electrons	Toryces of parental harmonic electrons					
	Symbols	g	$\tilde{b}_1$	$\tilde{\beta}_{1t}$	$\tilde{f}$, Hz	$\tilde{E}^-$, MeV
$e_{H,0,0}^{-H,1,0}$	$-\tilde{E}_{H,0,0}^{-H,1,0}$	-1.5000000	111.07915	0.99099741	4.119×10^{19}	0.340666
	$+\breve{\tilde{E}}_{H,0,0}^{-H,1,0}$	1.5000000	-110.07915i	1.00908437	-4.119×10^{19}	
$e_{H,1,0}^{-H,2,0}$	$-\tilde{E}_{H,1,0}^{-H,2,0}$	-0.8333333	213.14152	0.99530828	1.545×10^{19}	0.127750
	$+\breve{\tilde{E}}_{H,1,0}^{-H,2,0}$	0.8333333	-212.14152i	1.00471383	-1.545×10^{19}	
$e_{H,2,0}^{-H,3,0}$	$-\breve{\tilde{E}}_{H,2,0}^{-H,3,0}$	-0.5833333	323.83150	0.99691197	8.237×10^{18}	0.068134
	$+\breve{\tilde{E}}_{H,2,0}^{-H,3,0}$	0.5833333	-322.83150i	1.00309759	-8.237×10^{19}	

18.1.3 Parental harmonic positrons – see Table 18.1.2.

Table 18.1.3. Compositions and properties of parental harmonic positrons of all spacetime levels L with the quantum states $m = 0$.

Parental positrons	Toryces of parental harmonic positrons					
	Symbols	b_1	$\beta_1 = \beta_{2t}$	β_{2r}	δ_2	g
$e_{H,0,0}^{+1}$	$E_{H,0,0}^{+1/2}$	0.66666667	0.8660254	-0.50000000	1.50000000	1.5000000
	$\breve{E}_{H,0,0}^{+2}$	0.33333333	1.7320508i	-2.00000000	3.00000000	-1.5000000
$e_{H,1,0}^{+1}$	$E_{H,1,0}^{+2/3}$	0.60000000	0.7453560	-0.66666667	1.66666667	0.8333333
	$\breve{E}_{H,1,0}^{+3/2}$	0.40000000	1.1180340i	-1.50000000	2.50000000	-0.8333333
$e_{H,2,0}^{+1}$	$E_{H,2,0}^{+3/4}$	0.57142857	0.6614378	-0.75000000	1.75000000	0.5833333
	$\breve{E}_{H,2,0}^{+4/3}$	0.42857143	0.8819171i	-1.33333333	2.33333333	-0.5833333
$e_{H,3,0}^{+1}$	$E_{H,3,0}^{+4/5}$	0.55555556	0.60000000	-0.80000000	1.80000000	0.4500000
	$\breve{E}_{H,3,0}^{+5/4}$	0.44444444	0.7500000i	-1.25000000	2.25000000	-0.4500000

18.1.4 Harmonic positons – see Table 18.1.4.

Table 18.1.4. Compositions and properties of harmonic positons emitted by their parental positrons in all spacetime levels L with the quantum states $m = 0$.

Emitted positons	Helyces of emitted harmonic positons					
	Symbols	g	$\tilde{b}_1$	$\tilde{\beta}_{1t}$	$\tilde{f}$, Hz	$\tilde{E}^+$, MeV
$e^{+H,1,0}_{H,0,0}$	$+\tilde{E}^{+H,1,0}_{H,0,0}$	1.5000000	-110.07915i	1.00016356	-4.119×10^{19}	0.340666
	$-\tilde{E}^{+H,1,0}_{H,0,0}$	-1.5000000	111.07915	0.99099741	4.119×10^{19}	
$e^{+H,2,0}_{H,1,0}$	$+\tilde{E}^{+H,2,0}_{H,1,0}$	0.8333333	-212.14152i	1.00004423	-1.545×10^{19}	0.127750
	$-\tilde{E}^{+H,2,0}_{H,1,0}$	-0.8333333	213.14152	0.99530828	1.545×10^{19}	
$e^{+H,3,0}_{H,2,0}$	$+\tilde{E}^{+H,3,0}_{H,2,0}$	0.5833333	-322.83150i	1.00001913	-8.237×10^{19}	0.068134
	$-\tilde{E}^{+H,3,0}_{H,2,0}$	-0.5833333	323.83150	0.99691197	8.237×10^{18}	

18.1.5 Analysis of calculated data for harmonic electons & positons - It follows from Tables 18.1.2 and 18.1.4:

- Real and imaginary helyces making up harmonic electons and positons of all spacetime levels are emitted in the same direction.
- The translational velocities of leading strings $\tilde{\beta}_{1t}$ of emitted helyces with negative and positive goldicities g are respectively slightly less and greater than velocity of light.
- The calculated frequencies of harmonic electons and positons $\tilde{f}$ are independent on the spacetime levels L and they are within the frequency range of the **gamma-ray electromagnetic waves**.

The following changes of parameters of trailing strings of constituent helyces of harmonic electons and positons occur as the harmonic excitation quantum state n increases:

- The relative radius $\tilde{b}_1$ increases.
- The frequency $\tilde{f}$ decreases.
- The energy $\tilde{E}^{\pm}$ of emitted harmonic electons and positons decreases.

18.2 Harmonic Singulatons & Ethertons

18.2.1 Parental harmonic singulatrons — See Table 18.2.1.

Table 18.2.1. Compositions and properties of parental harmonic singulatrons of all spacetime levels L with the quantum states $m = 0$.

Parental singulatrons	\multicolumn{6}{c}{Toryces of parental harmonic singulatrons}
	Symbols	b_1	$\beta_1 = \beta_{2t}$	β_{2r}	δ_2	g
$\breve{a}^0_{H,0,0}$	$\breve{A}^{-2}_{H,0,0}$	-1.000000	$-1.73205081i$	2.000000	-1.000000	1.5000000
	$\breve{A}^{+2}_{H,0,0}$	0.333333	$1.73205081i$	-2.000000	3.000000	-1.5000000
$\breve{a}^0_{H,1,0}$	$\breve{A}^{-3}_{H,1,0}$	-0.500000	$-2.82842712i$	3.000000	-2.000000	2.6666667
	$\breve{A}^{+3}_{H,1,0}$	-0.250000	$2.82842712i$	-3.000000	4.000000	-2.6666667
$\breve{a}^0_{H,2,0}$	$\breve{A}^{-3}_{H,2,0}$	-0.333333	$-3.87298335i$	4.000000	-3.000000	3.7500000
	$\breve{A}^{+3}_{H,2,0}$	0.200000	$3.87298335i$	-4.000000	5.000000	-3.7500000
$\breve{a}^0_{H,3,0}$	$\breve{A}^{-4}_{H,3,0}$	-0.250000	$-4.89897949i$	5.000000	-4.000000	4.8000000
	$\breve{A}^{+4}_{H,3,0}$	0.166667	$4.89897949i$	-5.000000	6.000000	-4.8000000

18.2.2 Emitted harmonic singulatons − See Table 18.2.2.

Table 18.2.2. Compositions and properties of harmonic singulatons emitted by their parental singulatrons in all spacetime levels L with the quantum states $m = 0$.

Emitted singulatons	Helyces of emitted harmonic singulatons					
	Symbols	g	$\widetilde{b}_1$	$\widetilde{\beta}_{1t}$	$\widetilde{f}$, Hz	$\widetilde{\widetilde{E}}^0$, GeV
$\widetilde{a}_{H,0,0}^{H,1,0}$	$+\widetilde{A}_{H,0,0}^{H,1,0}$	1.5000000	0.500000142	-0.999998868	5.080×10^{22}	0.420151
	$-\widetilde{A}_{H,0,0}^{H,1,0}$	-1.5000000	0.499999858	-1.000001133	-5.080×10^{22}	
$\widetilde{a}_{H,1,0}^{H,2,0}$	$+\widetilde{A}_{H,1,0}^{H,2,0}$	2.6666667	0.500000122	-0.999999024	9.030×10^{22}	0.746936
	$-\widetilde{A}_{H,1,0}^{H,2,0}$	-2.6666667	0.499999878	-1.000000977	-9.030×10^{22}	
$\widetilde{a}_{H,2,0}^{H,3,0}$	$+\widetilde{A}_{H,2,0}^{H,3,0}$	3.7500000	0.500000115	-0.999999083	1.270×10^{23}	1.050038
	$-\widetilde{A}_{H,2,0}^{H,3,0}$	-3.7500000	0.499999885	-1.000000917	-1.270×10^{23}	

18.2.3 Parental harmonic ethertrons − See Table 18.2.3.

Table 18.2.3. Compositions and properties of parental harmonic ethertrons of all spacetime levels L with the quantum states $m = 0$.

Parental ethertrons	Toryces of parental harmonic ethertrons					
	Symbols	b_1	$\beta_1 = \beta_{2t}$	β_{2r}	δ_2	g
$a_{H,0,0}^0$	$A_{H,0,0}^{-1/2}$	2.00000000	0.86602540	0.50000000	0.50000000	-1.50000000
	$A_{H,0,0}^{+1/2}$	0.66666667	0.86602540	-0.50000000	1.50000000	1.50000000
$a_{H,1,0}^0$	$A_{H,1,0}^{-1/3}$	1.50000000	0.94280904	0.33333333	0.66666667	-2.66666667
	$A_{H,1,0}^{+1/3}$	0.75000000	0.94280904	-0.33333333	1.33333333	2.66666667
$a_{H,2,0}^0$	$A_{H,2,0}^{-1/3}$	1.33333333	0.96824584	0.25000000	0.75000000	-3.75000000
	$A_{H,2,0}^{+1/3}$	0.80000000	0.96824584	-0.25000000	1.25000000	3.75000000
$a_{H,3,0}^0$	$A_{H,3,0}^{-1/4}$	1.25000000	0.97979690	0.20000000	0.80000000	-4.80000000
	$A_{H,3,0}^{+1/4}$	0.83333333	0.97979690	-0.20000000	1.20000000	4.80000000

18.2.4 Emitted harmonic ethertons – See Table 18.2.4.

Table 18.2.4. Compositions and properties of harmonic ethertons emitted by their parental ethertons in all spacetime levels L with the quantum states $m = 0$.

Emitted ethertons	Helyces of emitted harmonic ethertons					
	Symbols	g	$\widetilde{b}_1$	$\widetilde{\beta}_{1t}$	$\widetilde{f}$, Hz	$\widetilde{E}^0$, GeV
$\widetilde{a}_{H,0,0}^{H,1,0}$	$-\widetilde{A}_{H,0,0}^{H,1,0}$	-1.5000000	0.499999858	-1.000001133	-5.080×10^{22}	0.420151
	$+\widetilde{A}_{H,0,0}^{H,1,0}$	1.5000000	0.500000142	-0.999998868	5.080×10^{22}	
$\widetilde{a}_{H,1,0}^{H,2,0}$	$-\widetilde{A}_{H,1,0}^{H,2,0}$	-2.6666667	0.499999878	-1.000000977	-9.030×10^{22}	0.746936
	$+\widetilde{A}_{H,1,0}^{H,2,0}$	2.6666667	0.500000122	-0.999999024	9.030×10^{22}	
$\widetilde{a}_{H,2,0}^{H,3,0}$	$-\widetilde{A}_{H,2,0}^{H,3,0}$	-3.7500000	0.499999885	-1.000000917	-1.270×10^{23}	1.050038
	$+\widetilde{A}_{H,2,0}^{H,3,0}$	3.7500000	0.500000115	-0.999999083	1.270×10^{23}	

18.2.5 Analysis of calculated data for harmonic singulatons & ethertons - It follows from Tables 18.2.2 and 18.2.4:

- Real and imaginary helyces making up harmonic singulatons and ethertons are emitted in the same direction.
- The translational velocities of leading strings $\widetilde{\beta}_{1t}$ of emitted helyces with positive and negative goldicities g are respectively slightly less or greater than velocity of light.
- The calculated frequencies of electons and positons $\widetilde{f}$ are independent on the spacetime levels L and they are at the upper end of frequency range of the ***gamma-ray electromagnetic waves***.

The following changes of parameters of trailing strings of constituent helyces of harmonic electons and positons occur as the harmonic excitation quantum state n increases:

- The relative radius $\widetilde{b}_1$ of helyces with negative and positive goldicities g respectively increase and decrease.
- The frequency $\widetilde{f}$ increases.
- The energy $\widetilde{E}^{\pm}$ of emitted harmonic electons and positons increases.

18.3 Oscillated Lambda Electrinos

18.3.1 Parental oscillated lambda electrons – See Table 18.3.1 and Fig. 11.1.

Table 18.3.1. Compositions and properties of parental oscillated lambda electrons of the ordinary matter $L1$ with $m = 2$, $n = 1$ and $b_1 = 37538.0$, $\breve{b}_1 = -37537.0$.

q	Q_q	Parental electrons	Symbols	$\beta_1 = \beta_{2t}$	β_{2r}	δ_2	g
				Toryces of parental oscillated lambda electrons			
0	1.00	$e_{2,1,0}^{-1}$	$E_{2,1,0}^{-}$	0.00729922	0.99997336	0.00002664	-5.328×10^{-5}
			$\breve{E}_{2,1,0}^{-}$	$-0.00729942i$	1.00002664	-0.00002664	5.328×10^{-5}
1	3.00	$e_{2,1,1}^{-1}$	$E_{2,1,1}^{-}$	0.00729922	0.99997336	0.00002664	-5.328×10^{-5}
			$\breve{E}_{2,1,1}^{-}$	$-0.00729942i$	1.00002664	-0.00002664	5.328×10^{-5}
2	205.5	$e_{2,1,2}^{-1}$	$E_{2,1,2}^{-}$	0.00729922	0.99997336	0.00002664	-5.328×10^{-5}
			$\breve{E}_{2,1,2}^{-}$	$-0.00729942i$	1.00002664	-0.00002664	5.328×10^{-5}
3	3519.1875	$e_{2,1,3}^{-1}$	$E_{2,1,3}^{-}$	0.00729922	0.99997336	0.00002664	-5.328×10^{-5}
			$\breve{E}_{2,1,3}^{-}$	$-0.00729942i$	1.00002664	-0.00002664	5.328×10^{-5}
4	35713.124	$e_{2,1,4}^{-1}$	$E_{2,1,4}^{-}$	0.00729922	0.99997336	0.00002664	-5.328×10^{-5}
			$\breve{E}_{2,1,4}^{-}$	$-0.00729942i$	1.00002664	-0.00002664	5.328×10^{-5}
5	258014.18	$e_{2,1,5}^{-1}$	$E_{2,1,5}^{-}$	0.00729922	0.99997336	0.00002664	-5.328×10^{-5}
			$\breve{E}_{2,1,5}^{-}$	$-0.00729942i$	1.00002664	-0.00002664	5.328×10^{-5}
6	1447851.7	$e_{2,1,6}^{-1}$	$E_{2,1,6}^{-}$	0.00729922	0.99997336	0.00002664	-5.328×10^{-5}
			$\breve{E}_{2,1,6}^{-}$	$-0.00729942i$	1.00002664	-0.00002664	5.328×10^{-5}
7	6642891.8	$e_{2,1,7}^{-1}$	$E_{2,1,7}^{-}$	0.00729922	0.99997336	0.00002664	-5.328×10^{-5}
			$\breve{E}_{2,1,7}^{-}$	$-0.00729942i$	1.00002664	-0.00002664	5.328×10^{-5}

18.3.2 Emitted oscillated lambda electrinos – See Table 18.3.2.

Table 18.3.2. Compositions and properties of oscillated lambda electrinos emitted by their parental electrons in the ordinary matter *L1* with $m = 2$, $n = 1$ and $b_1 = 37538.0$, $\breve{b}_1 = -37537.0$.

q_k - q_j	Emitted electrinos	Helyces of emitted oscillated lambda electrinos					
		Symbols	$\widetilde{b}_1$	$\widetilde{\beta}_{1t}$	$\widetilde{\beta}_{1r}$	$\widetilde{f}$, Hz	$\widetilde{E}_q^-$, MeV
1 - 0	e-trino $\widetilde{e}_{2,1,0}^{-2,1,1}$	$\widetilde{E}_{2,1,0}^{-2,1,1}$	37544.74519	0.99994673	0.00729876	-6.583×10^{15}	0.000054
		$\widetilde{\breve{E}}_{2,1,0}^{-2,1,1}$	-37543.74519	1.00005327	-0.00729857i	6.583×10^{15}	
2 - 1	3e-trino $\widetilde{e}_{2,1,1}^{-2,1,2}$	$\widetilde{E}_{2,1,1}^{-2,1,2}$	1728.777000	0.99884211	0.03402781	-6.665×10^{17}	0.005517
		$\widetilde{\breve{E}}_{2,1,1}^{-2,1,2}$	-1727.777000	1.00115655	-0.03400813i	6.665×10^{17}	
3 - 2	μ-trino $\widetilde{e}_{2,1,2}^{-2,1,3}$	$\widetilde{E}_{2,1,2}^{-2,1,3}$	268.6370000	0.99251302	0.08652612	-1.091×10^{19}	0.090218
		$\widetilde{\breve{E}}_{2,1,2}^{-2,1,3}$	-267.6370000	1.00743134	-0.08620402i	1.091×10^{19}	
4 - 3	τ-trino $\widetilde{e}_{2,1,3}^{-2,1,4}$	$\widetilde{E}_{2,1,3}^{-2,1,4}$	59.39348000	0.96543569	0.18585928	-1.060×10^{20}	0.876516
		$\widetilde{\breve{E}}_{2,1,3}^{-2,1,4}$	-58.39348000	1.03340888	-0.18272999i	1.060×10^{20}	
5 - 4	v-trino $\widetilde{e}_{2,1,4}^{-2,1,5}$	$\widetilde{E}_{2,1,4}^{-2,1,5}$	16.75087500	0.86784750	0.36195009	-7.317×10^{20}	6.052374
		$\widetilde{\breve{E}}_{2,1,4}^{-2,1,5}$	-15.75087500	1.11662022	-0.34034226i	7.317×10^{20}	
6 - 5	ρ-trino $\widetilde{e}_{2,1,5}^{-2,1,6}$	$\widetilde{E}_{2,1,5}^{-2,1,6}$	5.83894500	0.48039929	0.67529267	-3.916×10^{21}	32.394558
		$\widetilde{\breve{E}}_{2,1,5}^{-2,1,6}$	4.83894500	1.33011899	-0.55963947i	3.916×10^{21}	

18.3.3 Parental oscillated lambda positrons – See Table 18.3.3.

Table 18.3.3. Compositions and properties of parental lambda positrons of the ordinary matter *L1* with $m = 2$, $n = 1$, and $b_1 = 0.50000666$, $\breve{b}_1 = 0.49999334$.

q	Q_q	Parental positrons	Toryces of parental oscillated lambda positrons				
			Symbols	$\beta_1 = \beta_{2t}$	β_{2r}	δ_2	g
0	1.00	$e^{+1}_{2,1,0}$	$E^+_{2,1,0}$	0.00729922	-0.99997336	1.99997336	-5.328×10^{-5}
			$\breve{E}^+_{2,1,0}$	0.00729942i	-1.00002664	2.00002664	5.328×10^{-5}
1	3.00	$e^{+1}_{2,1,1}$	$E^+_{2,1,1}$	0.00729922	-0.99997336	1.99997336	-5.328×10^{-5}
			$\breve{E}^+_{2,1,1}$	0.00729942i	-1.00002664	2.00002664	5.328×10^{-5}
2	205.5	$e^{+1}_{2,1,2}$	$E^+_{2,1,2}$	0.00729922	-0.99997336	1.99997336	-5.328×10^{-5}
			$\breve{E}^+_{2,1,2}$	0.00729942i	-1.00002664	2.00002664	5.328×10^{-5}
3	3519.1875	$e^{+1}_{2,1,3}$	$E^+_{2,1,3}$	0.00729922	-0.99997336	1.99997336	-5.328×10^{-5}
			$\breve{E}^+_{2,1,3}$	0.00729942i	-1.00002664	2.00002664	5.328×10^{-5}
4	35713.124	$e^{+1}_{2,1,4}$	$E^+_{2,1,4}$	0.00729922	-0.99997336	1.99997336	-5.328×10^{-5}
			$\breve{E}^+_{2,1,4}$	0.00729942i	-1.00002664	2.00002664	5.328×10^{-5}
5	258014.18	$e^{+1}_{2,1,5}$	$E^+_{2,1,5}$	0.00729922	-0.99997336	1.99997336	-5.328×10^{-5}
			$\breve{E}^+_{2,1,5}$	0.00729942i	-1.00002664	2.00002664	5.328×10^{-5}
6	1447851.7	$e^{+1}_{2,1,6}$	$E^+_{2,1,6}$	0.00729922	-0.99997336	1.99997336	-5.328×10^{-5}
			$\breve{E}^+_{2,1,6}$	0.00729942i	-1.00002664	2.00002664	5.328×10^{-5}

18.3.4 Emitted oscillated lambda positrinos – See Table 18.3.4.

Table 18.3.4. Compositions and properties of oscillated lambda positrinos emitted by their parental positrons in the ordinary matter *L1* with $m = 2$, $n = 1$ and

$$\breve{b}_1 = 0.50000666 \ , \ \breve{b}_1 = 0.49999334 \ .$$

$q_k - q_j$	Emitted positrinos	\multicolumn{6}{c}{Helyces of emitted oscillated lambda positrinos}
		Symbols	$\breve{b}_1$	$\breve{\beta}_{1t}$	$\breve{\beta}_{1r}$	$\breve{f}$, Hz	$\breve{E}_q^+$, MeV
1 - 0	*e-trino* $\widetilde{e}_{2,1,0}^{+2,1,1}$	$\widetilde{\breve{E}}_{2,1,0}^{+2,1,1}$	-37543.74519	1.00005327	-0.00729857i	6.583×10^{15}	0.000054
		$\breve{E}_{2,1,0}^{+2,1,1}$	37544.74519	0.99994673	0.00729876	-6.583×10^{15}	
2 - 1	*3e-trino* $\widetilde{e}_{2,1,1}^{+2,1,2}$	$\widetilde{\breve{E}}_{2,1,1}^{+2,1,2}$	-1727.777000	1.00115655	-0.03400813i	6.665×10^{17}	0.005517
		$\breve{E}_{2,1,1}^{+2,1,2}$	1728.777000	0.99884211	0.03402781	-6.665×10^{17}	
3 - 2	*μ-trino* $\widetilde{e}_{2,1,2}^{+2,1,3}$	$\widetilde{\breve{E}}_{2,1,2}^{+2,1,3}$	-267.6370000	1.00743134	-0.08620402i	1.091×10^{19}	0.090218
		$\breve{E}_{2,1,2}^{+2,1,3}$	268.6370000	0.99251302	0.08652612	-1.091×10^{19}	
4 - 3	*τ-trino* $\widetilde{e}_{2,1,3}^{+2,1,4}$	$\breve{E}_{2,1,3}^{+2,1,4}$	-58.39348000	1.03340888	-0.18272999i	1.060×10^{20}	0. 876516
		$\widetilde{\breve{E}}_{2,1,3}^{+2,1,4}$	59.39348000	0.96543569	0.18585928	-1.060×10^{20}	
5 - 4	*v-trino* $\widetilde{e}_{2,1,4}^{+2,1,5}$	$\widetilde{\breve{E}}_{2,1,4}^{+2,1,5}$	-15.75087500	1.11662022	-0.34034226i	7.317×10^{20}	6.052374
		$\breve{E}_{2,1,4}^{+2,1,5}$	16.75087500	0.86784750	0.36195009	-7.317×10^{20}	
6 - 5	*ρ-trino* $\widetilde{e}_{2,1,5}^{+2,1,6}$	$\widetilde{\breve{E}}_{2,1,5}^{+2,1,6}$	4.83894500	1.33011899	-0.55963947i	3.916×10^{21}	32.394558
		$\breve{E}_{2,1,5}^{+2,1,6}$	5.83894500	0.48039929	0.67529267	-3.916×10^{21}	

18.3.5. Analysis of calculated properties of oscillated lambda electrinos & positrinos – It follows from Tables 18.3.2 and 18.3.4:

- Real and imaginary helyces making up oscillated lambda electrinos and positrinos are emitted in the same direction.

- The translational velocities of leading strings $\widetilde{\beta}_{1t}$ of emitted real helyces making up lambda electrinos are less than velocity of light, while these velocities of emitted imaginary helyces are greater than velocity of light.
- The translational velocities of leading strings $\widetilde{\beta}_{1t}$ of emitted real helyces making up lambda positrinos are greater than velocity of light, while these velocities of emitted imaginary helyces are less than velocity of light.

The following changes of parameters of electrinos and positrinos occur as the oscillation quantum states q increase:

- The relative radii of leading strings $\widetilde{b}_1$ of their constituent helyces decrease
- The frequencies $\widetilde{f}$ of their constituent helyces increase
- The energies $\widetilde{E}_q$ of emitted oscillated lambda electrinos and positrinos increase.

18.4 Oscillated Lambda Singulatrinos & Ethertrinos

18.4.1 Parental oscillated lambda singulatrons – See Table 18.4.1.

Table 18.4.1. Compositions and properties of parental oscillated lambda singulatrons of the ordinary matter *L1* with the quantum states $m = n = 1$ and $\breve{b}_1^- = -0.00366300$, $\breve{b}_1^+ = +0.00363636$.

q	Q_q	Parental singulatrons	Symbols	$\beta_1 = \beta_{2t}$	β_{2r}	δ_2	g
				Toryces of parental oscillated lambda singulatrons			
0	1.00	$\breve{a}_{1,1,0}^0$	$\tilde{\tilde{A}}_{1,1,0}^-$	-273.998175i	274.000000	-273.000273	273.9964
			$\tilde{\tilde{A}}_{1,1,0}^+$	-273.998175i	274.000000	-275.000275	-273.9964
1	3.00	$\breve{a}_{1,1,1}^0$	$\tilde{\tilde{A}}_{1,1,1}^-$	-273.998175i	274.000000	-273.000273	273.9964
			$\tilde{\tilde{A}}_{1,1,1}^+$	-273.998175i	274.000000	--275.000275	-273.9964
2	205.5	$\breve{a}_{1,1,2}^0$	$\tilde{\tilde{A}}_{1,1,2}^-$	-273.998175i	274.000000	-273.000273	273.9964
			$\tilde{\tilde{A}}_{1,1,2}^+$	-273.998175i	274.000000	-275.000275	-273.9964
3	3519.1875	$\breve{a}_{1,1,3}^0$	$\tilde{\tilde{A}}_{1,1,3}^-$	-273.998175i	274.000000	-273.000273	273.9964
			$\tilde{\tilde{A}}_{1,1,3}^+$	-273.998175i	274.000000	-275.000275	-273.9964
4	35713.124	$\breve{a}_{1,1,4}^0$	$\tilde{\tilde{A}}_{1,1,4}^-$	-273.998175i	274.000000	-273.000273	273.9964
			$\tilde{\tilde{A}}_{1,1,4}^+$	-273.998175i	274.000000	-275.000275	-273.9964
5	258014.18	$\breve{a}_{1,1,5}^0$	$\tilde{\tilde{A}}_{1,1,5}^-$	-273.998175i	274.000000	-273.000273	273.9964
			$\tilde{\tilde{A}}_{1,1,5}^+$	-273.998175i	274.000000	-275.000275	-273.9964
6	1447851.7	$\breve{a}_{1,1,6}^0$	$\tilde{\tilde{A}}_{1,1,6}^-$	-273.998175i	274.000000	-273.000273	273.9964
			$\tilde{\tilde{A}}_{1,1,6}^+$	-273.998175i	274.000000	-275.000275	-273.9964

18.4.2 Emitted oscillated lambda singulatrinos – See Table 18.4.2.

Table 18.4.2. Compositions and properties of oscillated lambda singulatrinos emitted by their parental singulatrons in the ordinary matter $L1$ with the quantum states $m = n = 1$ and $\breve{b}_1^- = -0.00366300$, $\breve{b}_1^+ = +0.00363636$.

Emitted singula-trinos	Helyces of emitted oscillated lambda singulatrinos					
	Symbols	g	$\widetilde{b}_1$	$\widetilde{\beta}_{1t}$	$\widetilde{f}$, Hz	$\widetilde{\widetilde{E}}^0$, GeV
$\widetilde{a}_{1,1,0}^{1,1,1}$	$+\widetilde{A}_{1,1,0}^{1,1,1}$	273.9964	0.46900682	-1.225416	3.386×10^{22}	0.28002
	$-\widetilde{A}_{1,1,0}^{1,1,1}$	-273.9964	-0.88341567	2.307659	-3.386×10^{22}	
$\widetilde{a}_{1,1,1}^{1,1,2}$	$+\widetilde{A}_{1,1,1}^{1,1,2}$	273.9964	0.00987878	-100.232063	3.428×10^{24}	28.3520
	$-\widetilde{A}_{1,1,1}^{1,1,2}$	-273.9964	-0.00987878	102.231965	-3.428×10^{24}	
$\widetilde{a}_{1,1,2}^{1,1,3}$	$+\widetilde{A}_{1,1,2}^{1,1,3}$	273.9964	0.00060372	-1655.386538	5.609×10^{25}	4.64×10^2
	$-\widetilde{A}_{1,1,2}^{1,1,3}$	-273.9964	-0.00060372	1657.386538	-5.609×10^{25}	
$\widetilde{a}_{1,1,3}^{1,1,4}$	$+\widetilde{A}_{1,1,3}^{1,1,4}$	273.9964	0.00006214	-16091.57938	5.450×10^{26}	4.51×10^3
	$-\widetilde{A}_{1,1,3}^{1,1,4}$	-273.9964	-0.00006214	16093.57938	-5.450×10^{26}	
$\widetilde{a}_{1,1,4}^{1,1,5}$	$+\widetilde{A}_{1,1,4}^{1,1,5}$	273.9964	0.00000900	-111118.7796	3.763×10^{27}	3.11×10^4
	$-\widetilde{A}_{1,1,4}^{1,1,5}$	-273.9964	-0.00000900	111120.7796	-3.763×10^{27}	
$\widetilde{a}_{1,1,5}^{1,1,6}$	$+\widetilde{A}_{1,1,5}^{1,1,6}$	273.9964	0.00060168	-594753.4984	2.014×10^{28}	1.67×10^5
	$-\widetilde{A}_{1,1,5}^{1,1,6}$	-273.9964	-0.00000168	594755.4984	-2.014×10^{28}	

18.4.3 Parental oscillated lambda ethertrons – See Table 18.4.3.

Table 18.4.3. Compositions and properties of parental oscillated lambda ethertrons of the ordinary matter $L1$ with the quantum states $m = n = 1$ and $b_1^- = 1.00366300$, $b_1^+ = 0.99636364$.

q	Q_q	Parental ethertrons	Symbols	$\beta_1 = \beta_{2t}$	β_{2r}	δ_2	g
0	1.00	$a_{1,1,0}^0$	$A_{1,1,0}^-$	0.99999334	0.00364964	0.996350369	-273.9964
			$A_{1,1,0}^+$	0.99999334	-0.00364964	1.003639631	273.9964
1	3.00	$a_{1,1,1}^0$	$A_{1,1,1}^-$	0.99999334	0.00364964	0.996350369	-273.9964
			$A_{1,1,1}^+$	0.99999334	-0.00364964	1.003639631	273.9964
2	205.5	$a_{1,1,2}^0$	$A_{1,1,2}^-$	0.99999334	0.00364964	0.996350369	-273.9964
			$A_{1,1,2}^+$	0.99999334	-0.00364964	1.003639631	273.9964
3	3519.1875	$a_{1,1,3}^0$	$A_{1,1,3}^-$	0.99999334	0.00364964	0.996350369	-273.9964
			$A_{1,1,3}^+$	0.99999334	-0.00364964	1.003639631	273.9964
4	35713.124	$a_{1,1,4}^0$	$A_{1,1,4}^-$	0.99999334	0.00364964	0.996350369	-821.989
			$A_{1,1,4}^+$	0.99999334	-0.00364964	1.003639631	-273.9964
5	258014.18	$a_{1,1,5}^0$	$A_{1,1,5}^-$	0.99999334	0.00364964	0.996350369	-273.9964
			$A_{1,1,5}^+$	0.99999334	-0.00364964	1.003639631	273.9964
6	1447851.7	$a_{1,1,6}^0$	$A_{1,1,6}^-$	0.99999334	0.00364964	0.996350369	-273.9964
			$A_{1,1,6}^+$	0.99999334	-0.00364964	1.003639631	273.9964

The header row "Toryces of parental oscillated lambda ethertrons" spans the columns Symbols, $\beta_1 = \beta_{2t}$, β_{2r}, δ_2, g.

18.4.4 Emitted oscillated lambda ethertrinos – See Table 18.4.4.

Table 18.4.4. Compositions and properties of oscillated lambda ethertrinos emitted by their parental ethertrons in the ordinary matter *L1* with the quantum states $m = n = 1$ and $b_1^- = 1.00366300$, $b_1^+ = 0.99636364$.

Emitted ethertrinos	Helyces of emitted oscillated lambda ethertrinos					
	Symbols	g	$\tilde{b}_1$	$\tilde{\beta}_{1t}$	$\tilde{f}$, Hz	$\tilde{\tilde{E}}^0$, GeV
$a_{1,1,0}^{1,1,1}$	$-\tilde{A}_{1,1,0}^{1,1,1}$	-273.9964	0.46900682	-1.225416	3.386×10^{22}	0.28002
	$+\tilde{A}_{1,1,0}^{1,1,1}$	273.9964	-0.88341567	2.307659	-3.386×10^{22}	
$a_{1,1,1}^{1,1,2}$	$-\tilde{A}_{1,1,1}^{1,1,2}$	-273.9964	0.00987878	-100.232063	3.428×10^{24}	28.3520
	$+\tilde{A}_{1,1,1}^{1,1,2}$	273.9964	-0.00987878	102.231965	-3.428×10^{24}	
$a_{1,1,2}^{1,1,3}$	$-\tilde{A}_{1,1,2}^{1,1,3}$	-273.9964	0.00060372	-1655.386538	5.609×10^{25}	4.64×10^2
	$+\tilde{A}_{1,1,2}^{1,1,3}$	273.9964	-0.00060372	1657.386538	-5.609×10^{25}	
$a_{1,1,3}^{1,1,4}$	$-\tilde{A}_{1,1,3}^{1,1,4}$	-273.9964	0.00006214	-16091.57938	5.450×10^{26}	4.51×10^3
	$+\tilde{A}_{1,1,3}^{1,1,4}$	273.9964	-0.00\ 006214	16093.57938	-5.450×10^{26}	
$a_{1,1,4}^{1,1,5}$	$-\tilde{A}_{1,1,4}^{1,1,5}$	-273.9964	0.00000900	-111118.7796	3.763×10^{27}	3.11×10^4
	$+\tilde{A}_{1,1,4}^{1,1,5}$	273.9964	-0.00000900	111120.7796	-3.763×10^{27}	
$a_{1,1,5}^{1,1,6}$	$-\tilde{A}_{1,1,5}^{1,1,6}$	-273.9964	0.00060168	-594753.4984	2.014×10^{28}	1.67×10^5
	$+\tilde{A}_{1,1,5}^{1,1,6}$	273.9964	-0.00000168	594755.4984	-2.014×10^{28}	

18.4.5 Analysis of calculated data for oscillated lambda singulatrinos & ethertrinos - It follows from Tables 18.4.2 and 18.4.4:

- Positive and negative helyces making up oscillated lambda singulatrinos and ethertrinos are emitted in opposite directions.
- The translational velocities of leading strings $\tilde{\beta}_{1t}$ of emitted helyces making up oscillated lambda singulatrinos and ethertrinos substantially greater than velocity of light.

The following changes of parameters of singulatrinos and ethertrinos occur as the oscillation quantum states q increase:

- The relative radii of leading strings $\tilde{b}_1$ of their constituent helyces decrease.
- The superluminal translational velocities of leading strings $\tilde{\beta}_{1t}$ of emitted helyces significantly increase.
- The frequency $\tilde{f}$ of their constituent helyces increase and they are above the upper end of frequency range of the ***gamma-ray electromagnetic waves.***
- The energies $\tilde{E}_q$ of emitted electrinos increase.

18.5 Oscillated Harmonic Electrinos & Positrinos

18.5.1 Parental oscillated harmonic electrons – See Table 18.5.1.

Table 18.5.1. Compositions and properties of parental oscillated harmonic electrons of all spacetime levels L with $m = n = 0$ and $b_1 = 2.0$, $\breve{b}_1 = -1.0$.

q	Q_q	Parental electrons	Toryces of parental oscillated harmonic electrons				
			Symbols	$\beta_1 = \beta_{2t}$	β_{2r}	δ_2	g
0	1.00	$e_{H,0,0}^{-1}$	$E_{H,0,0}^-$	0.86602540	0.50000000	0.50000000	-1.50000000
			$\breve{E}_{H,0,0}^-$	-1.73205081i	2.00000000	-1.00000000	1.50000000
1	3.00	$e_{H,0,1}^{-1}$	$E_{H,0,1}^-$	0.86602540	0.50000000	0.50000000	-1.50000000
			$\breve{E}_{H,0,1}^-$	-1.73205081i	2.00000000	-1.00000000	1.50000000
2	205.5	$e_{H,0,2}^{-1}$	$E_{H,0,2}^-$	0.86602540	0.50000000	0.50000000	-1.50000000
			$\breve{E}_{H,0,2}^-$	-1.73205081i	2.00000000	-1.00000000	1.50000000
3	3519.1875	$e_{H,0,3}^{-1}$	$E_{H,0,3}^-$	0.86602540	0.50000000	0.50000000	-1.50000000
			$\breve{E}_{H,0,3}^-$	-1.73205081i	2.00000000	-1.00000000	1.50000000

18.5.2 Emitted oscillated harmonic electrinos – See Table 18.5.2.

Table 18.5.2. Compositions and properties of oscillated harmonic electrinos emitted by their parental electrons in all spacetime levels L with $m = n = 0$ and $b_1 = 2.0$, $\breve{b}_1 = -1.0$.

$q_k - q_j$	Emitted electrinos	\multicolumn{6}{c}{Helyces of emitted oscillated harmonic electrinos}
		Symbols	$\widetilde{b}_1$	$\widetilde{\beta}_{1t}$	$\widetilde{\beta}_{1r}$	$\widetilde{f}$, Hz	$\widetilde{E}_q^-$, MeV
1 - 0	*e-trino* $\widetilde{e}_{H,1,0}^{-H,1,1}$	$\widetilde{E}_{H,0,0}^{-H,0,1}$	0.50000094	-0.99999626	-0.00273643	-1.853×10^{20}	-1.532
		$\breve{\widetilde{E}}_{H.0,0}^{-H,0,1}$	0.49999906	-1.00000374	-0.00273642	1.853×10^{20}	
2 - 1	*3e-trino* $\widetilde{e}_{H,0,1}^{-H,0,2}$	$\widetilde{E}_{H,0,1}^{-H,0,2}$	0.50958903	-0.96236563	-0.28238535	-1.877×10^{22}	-155.216
		$\breve{\widetilde{E}}_{H.0,1}^{-H,0,2}$	0.49041097	-1.03410610	-0.27175796	1.877×10^{22}	
3 - 2	*μ-trino* $\widetilde{e}_{H,0,2}^{-H,0,3}$	$\widetilde{E}_{H,0,2}^{-H,0,3}$	0.89055716	$-372.409117i$	-8.07550932	-3.071×10^{23}	-2539.934
		$\breve{\widetilde{E}}_{H.0,2}^{-H,0,3}$	0.10944285	$-372.497985i$	0.99241998	3.071×10^{23}	

18.5.3 Parental oscillated harmonic positrons – See Table 18.5.3.

Table 18.5.3. Compositions and properties of parental oscillated harmonic positrons of all spacetime levels L with $m = n = 0$ and $b_1 = \frac{2}{3}$, $\breve{b}_1 = \frac{1}{3}$.

q	Q_q	Parental positrons	\multicolumn{5}{c}{Toryces of parental oscillated harmonic positrons}
			Symbols	$\beta_1 = \beta_{2t}$	β_{2r}	δ_2	g
0	1.00	$e_{H,0,0}^{+1}$	$E_{H,0,0}^+$	0.86602540	-0.50000000	1.50000000	1.50000000
			$\breve{E}_{H,0,0}^+$	$1.73205081i$	-2.00000000	3.00000000	-1.50000000
1	3.00	$e_{H,0,1}^{+1}$	$E_{H,0,1}^+$	0.86602540	-0.50000000	1.50000000	1.50000000
			$\breve{E}_{H,0,1}^+$	$1.73205081i$	-2.00000000	3.00000000	-1.50000000
2	205.5	$e_{H,0,2}^{+1}$	$E_{H,0,2}^+$	0.86602540	-0.50000000	1.50000000	1.50000000
			$\breve{E}_{H,0,2}^+$	$1.73205081i$	-2.00000000	3.00000000	-1.50000000
3	3519.1875	$e_{H,0,3}^{+1}$	$E_{H,0,3}^+$	0.86602540	-0.50000000	1.50000000	1.50000000
			$\breve{E}_{H,0,3}^+$	$1.73205081i$	-2.00000000	3.00000000	-1.50000000

18.5.4 Emitted oscillated harmonic positrinos – See Table 18.5.4.

Table 18.5.4. Compositions and properties of oscillated harmonic positrinos emitted by their parental positrons in all spacetime levels L with $m = n = 0$ and

$$b_1 = \tfrac{2}{3}, \ \widetilde{b}_1 = \tfrac{1}{3}.$$

$q_k\text{-}q_j$	Emitted positrinos	Symbols	g	$\widetilde{b}_1$	$\widetilde{\beta}_{1t}$	$\widetilde{f}$, Hz	$\widetilde{E}_q^+$, MeV
1 – 0	*e-trino* $\widetilde{e}_{H,0,0}^{+H,0,1}$	$\widetilde{\widetilde{E}}_{H,0,0}^{+H,0,1}$	1.50000000	0.49999906	-1.00000374	1.853×10^{20}	-1.532
		$\widetilde{E}_{H,0,0}^{+H,0,1}$	-1.50000000	0.50000094	-0.99999626	-1.853×10^{20}	
2 – 1	*3e-trino* $\widetilde{e}_{H,0,1}^{+H,0,2}$	$\widetilde{\widetilde{E}}_{H,0,1}^{+H,0,2}$	1.50000000	0.49041097	-1.03410610	1.877×10^{22}	-155.216
		$\widetilde{E}_{H,0,1}^{+H,0,2}$	-1.50000000	0.50958903	-0.96236563	-1.877×10^{22}	
3 – 2	*μ-trino* $\widetilde{e}_{H,0,2}^{+H,0,3}$	$\widetilde{\widetilde{E}}_{H,0,2}^{+H,0,3}$	1.50000000	0.10944285	$-372.497985i$	3.071×10^{23}	-2539.934
		$\widetilde{E}_{H,0,2}^{+H,0,3}$	-1.50000000	0.89055716	$-372.409117i$	-3.071×10^{23}	

The header spanning cell above columns g, $\widetilde{b}_1$, $\widetilde{\beta}_{1t}$, $\widetilde{f}$, $\widetilde{E}_q^+$ reads: **Helyces of emitted oscillated harmonic positrinos**

18.5.5 Analysis of calculated properties of oscillated harmonic electrinos & positrinos
- It follows from Tables 18.5.2 and 18.5.4:

- Positive and negative helyces making up oscillated harmonic electrinos and positrinos are emitted in the same direction.
- The translational velocities of leading strings $\widetilde{\beta}_{1t}$ of emitted helyces making up oscillated harmonic electrinos and positrinos with negative and positive goldicities g are respectively slightly less and greater than velocity of light.

The following changes of parameters of electrinos and positrinos making up harmonic *e-trinos* and *3e-trinos* occur as the oscillation quantum states q increase:

- The translational velocities of leading strings $\widetilde{\beta}_{1t}$ of their constituent helyces increase while the real and imaginary helyces making up electrinos and positrinos are emitted in the same direction.

- The translational velocities of leading strings $\tilde{\beta}_{1t}$ of emitted real helyces are slightly less than velocity of light, while these velocities of emitted imaginary helyces are slightly greater than velocity of light.
- The difference between velocities of real and imaginary helyces increases.
- The relative radii of leading strings $\tilde{b}_1$ of their constituent helyces decrease.
- The frequencies $\tilde{f}$ of their constituent helyces increase.
- The energies $\tilde{E}_q$ of emitted electrinos and positrinos increase.

Notably, the emitted helyces making up positive and negative *μ-trinos*, and higher level *trinos* are involved in turbulent vortex motions with their velocities expressed by real, imaginary, subluminal and superluminal numbers.

18.6 Oscillated Harmonic Singulatrons & Ethertrons

18.6.1 Parental oscillated harmonic singulatrons – See Table 18.6.1.

Table 18.6.1. Compositions and properties of parental oscillated harmonic singulatrons of all spacetime levels L with $m = n = 0$ and

$$\breve{b}_1^- = -1, \, \breve{b}_1^+ = \tfrac{1}{3} \, .$$

q	Q_q	Parental singulatrons	Toryces of parental oscillated harmonic singulatrons				
			Symbols	$\beta_1 = \beta_{2t}$	β_{2r}	δ_2	g
0	1.00	$\breve{a}_{0,0,0}^{0}$	$\tilde{A}_{H,0,0}^{-}$	-1.73205081i	274.000000	-1.000000	1.500000
			$\tilde{A}_{0,0,0}^{+}$	-1.73205081i	-274.000000	3.000000	-1.500000
1	3.00	$\breve{a}_{0,0,1}^{0}$	$\tilde{A}_{H,0,1}^{-}$	-1.73205081i	274.000000	-1.000000	1.500000
			$\tilde{A}_{0,0,1}^{+}$	1.73205081i	-274.000000	3.000000	-1.500000
2	205.5	$\breve{a}_{0,0,2}^{0}$	$\tilde{A}_{H,0,2}^{-}$	-1.73205081i	274.000000	-1.000000	1.500000
			$\tilde{A}_{0,0,2}^{+}$	1.73205081i	-274.000000	3.000000	-1.500000
3	3519.1875	$\breve{a}_{0,0,3}^{0}$	$\tilde{A}_{H,0,3}^{-}$	-1.73205081i	274.000000	-1.000000	1.500000
			$\tilde{A}_{0,0,3}^{+}$	1.73205081i	-274.000000	3.000000	-1.500000

18.6.2 Emitted oscillated harmonic singulatrinos – See Table 18.6.2.

Table 18.6.2. Compositions and properties of oscillated harmonic singulatrinos emitted by their parental singulatrons in all spacetime levels L with $m = n = 0$ and $\breve{b}_1^- = -1$, $\breve{b}_1^+ = \frac{1}{3}$.

Emitted singula-trinons	Helyces of emitted oscillated harmonic singulatrinos					
	Symbols	g	$\tilde{b}_1$	$\tilde{\beta}_{1t}$	$\tilde{f}$, Hz	$\tilde{E}^0$, GeV
$\breve{a}_{0,0,0}^{0,0,1}$	$+\tilde{A}_{0,0,0}^{0,0,1}$	1.500000	0.49999906	-1.00000374	1.853×10^{20}	0.001532
	$-\tilde{A}_{0,0,0}^{0,0,1}$	-1.500000	0.50000094	-0.99999626	-1.853×10^{20}	
$\breve{a}_{0,0,1}^{0,0,2}$	$+\tilde{A}_{0,0,1}^{0,0,2}$	1.500000	0.49041097	-1.03910610	1.877×10^{22}	0.155216
	$-\tilde{A}_{0,0,1}^{0,0,1}$	-1.500000	0.50958903	-0.95236563	-1.877×10^{22}	
$\breve{a}_{0,0,2}^{0,0,3}$	$+\tilde{A}_{0,0,2}^{0,0,3}$	1.500000	0.10944285	-372.497985	3.071×10^{23}	2.539934
	$-\tilde{A}_{0,0,2}^{0,0,3}$	-1.500000	0.89055715	-372.409117	-3.071×10^{23}	

18.6.3 Parental oscillated harmonic ethertrons – See Table 18.6.3.

Table 18.6.3. Compositions and properties of parental oscillated harmonic ethertrons of all spacetime levels L with $m = n = 0$ and $b_1^- = 2$, $b_1^+ = \frac{2}{3}$.

q	Q_q	Parental ethertrons	Toryces of parental oscillated harmonic ethertrons				
			Symbols	$\beta_1 = \beta_{2t}$	β_{2r}	δ_2	g
0	1.00	$a_{1,1,0}^0$	$A_{1,1,0}^-$	0.99999334	0.00364964	0.996350369	-273.9964
			$A_{1,1,0}^+$	0.99999334	-0.00364964	1.003639631	273.9964
1	3.00	$a_{1,1,1}^0$	$A_{1,1,1}^-$	0.99999334	0.00364964	0.996350369	-273.9964
			$A_{1,1,1}^+$	0.99999334	-0.00364964	1.003639631	273.9964
2	205.5	$a_{1,1,2}^0$	$A_{1,1,2}^-$	0.99999334	0.00364964	0.996350369	-273.9964
			$A_{1,1,2}^+$	0.99999334	-0.00364964	1.003639631	273.9964
3	3519.1875	$a_{1,1,3}^0$	$A_{1,1,3}^-$	0.99999334	0.00364964	0.996350369	-273.9964
			$A_{1,1,3}^+$	0.99999334	-0.00364964	1.003639631	273.9964

18.6.4 Emitted oscillated harmonic ethertons – See Table 18.6.4.

Table 18.4.4. Compositions and properties of oscillated harmonic ethertons emitted by their parental ethertrons in all spacetime levels L with $m = n = 0$ and

$$b_1^- = 2 \,,\, b_1^+ = \tfrac{2}{3}\,.$$

Emitted ethertons	Helyces of emitted oscillated harmonic ethertons					
	Symbols	g	$\widetilde{b}_1$	$\widetilde{\beta}_{1t}$	$\widetilde{f}$, Hz	$\widetilde{\widetilde{E}}^0$, GeV
$a_{0,0,0}^{0,0,1}$	$- A_{0,0,0}^{0,0,1}$	-273.9964	0.50000094	-0.99999626	3.386×10^{22}	0.001532
	$+ A_{0,0,0}^{0,0,1}$	273.9964	0.49999996	-1.00000374	-3.386×10^{22}	
$a_{0,0,1}^{0,0,2}$	$- A_{0,0,1}^{0,0,2}$	-273.9964	0.50958903	-0.96236563	3.428×10^{24}	0.155216
	$+ A_{0,0,1}^{0,0,1}$	273.9964	0.49041097	-1.03910610	-3.428×10^{24}	
$a_{0,0,2}^{0,0,3}$	$- A_{0,0,2}^{0,0,3}$	-273.9964	0.89055715	-372.409117i	5.609×10^{25}	2.539934
	$+ A_{0,0,2}^{0,0,3}$	273.9964	0.10944285	-372.497985i	-5.609×10^{25}	

18.6.5 Analysis of calculated properties of oscillated harmonic singulatrinos & ethertrinos -

It follows from Tables 18.6.2 and 18.6.4:

The following changes of parameters of oscillated harmonic singulatrinos and ethertrinos making up harmonic *e-trinos* and *3e-trinos* occur as the oscillation quantum states q increase:

- The translational velocities of leading strings $\widetilde{\beta}_{1t}$ of their constituent helyces increase while the real and imaginary helyces making up electrinos and positrinos are emitted in the same direction
- The translational velocities of leading strings $\widetilde{\beta}_{1t}$ of emitted real helyces are slightly less than velocity of light, while these velocities of emitted imaginary helyces are slightly greater than velocity of light. The difference between velocities of real and imaginary helyces increases.
- The relative radii of leading strings $\widetilde{b}_1$ of their constituent helyces decrease
- The frequencies $\widetilde{f}$ of their constituent helyces increase
- The energies $\widetilde{E}_q$ of emitted electrinos and positrinos increase.

Notably, the emitted helyces making up positive and negative *μ-trinos, τ-trinos, ν-trinos,* and *ρ-trinos* are involved in turbulent vortex motions with their velocities expressed by real, imaginary, subluminal and superluminal numbers.

MAIN SUMMARY

Directions of propagation of real and imaginary helyces making up harmonic electons and positons are the same and their velocities are equal to the velocity of light for all spacetime levels of the Multiverse (Tables 18.1.2 and 18.1.4).

Directions of propagation of real and imaginary helyces making up harmonic singulatons and ethertons are the same and their velocities are equal to the velocity of light for all spacetime levels of the Multiverse (Tables 18.2.2 and 18.2.4).

Directions of propagation of real and imaginary helyces making up oscillated lambda electrinos and positrinos and are the same and their velocities are equal to the velocity of light for all oscillation quantum states (Tables 18.3.2 and 18.3.4).

Real and imaginary helyces making up oscillated lambda singulatrinos and ethertrinos propagate in opposite directions and their velocities are superluminal for all oscillation quantum states (Tables 18.4.2 and 18.4.4).

PHYSICAL & SPACETIME CONSTANTS

Physical Constants (NIST CODATA)

Names		Values
Elementary charge	e_0	$1.602\ 176\ 565 \times 10^{-19}$ C
Electric constant	ε_0	$8.854\ 187\ 817 \times 10^{-12}$ C^2/N/m^2
Electron mass	m_e	$9.109\ 382\ 91 \times 10^{-31}$ kg
Proton relative mass	m_p / m_e	$1836.152\ 671\ 95$
Neutron relative mass	m_n / m_e	$1838.683\ 660\ 08$
Muon relative mass	m_μ / m_e	$206.768\ 284\ 26$
Tau relative mass	m_τ / m_e	$3477.151\ 011\ 54$
Newtonian constant of gravitation	G	$6.673\ 84 \times 10^{-11}$ m^3/kg/s^2
Planck constant	h	$6.626\ 069\ 573 \times 10^{-34}$ J s
Inverse fine-structure constant	α^{-1}	$137.035\ 999\ 074$
Bohr magneton	μ_B	$9.274\ 009\ 68 \times 10^{-24}$ J/T
Muon magneton	μ_μ	$4.485\ 218\ 58 \times 10^{-26}$ J/T
Tau magneton	μ_τ	$2.666\ 876\ 44 \times 10^{-24}$ J/T
Electron magnetic moment to Bohr magneton ratio	μ_e / μ_B	$-1.001\ 159\ 652\ 180\ 76$
Muon magnetic moment to muon magneton ratio	μ_μ / μ_μ	$-1.001\ 165\ 923$
Tau magnetic moment to tau magneton ratio	μ_τ / μ_τ	$-0.987\ 629\ 486$
Nuclear magneton	μ_N	$5.050\ 783\ 54 \times 10^{-27}$ J/T
Proton magnetic moment to nuclear magneton ratio	μ_p / μ_N	$2.792\ 847\ 356$
Neutron magnetic moment to nuclear magneton ratio	μ_n / μ_N	$-1.913\ 042\ 72$

Spacetime Constants

Names		Values
Speed of light in vacuum	c	$2.997\,924\,58 \times 10^{+08}$ m/s
Classical electron radius	r_e	$2.817\,940\,323 \times 10^{-15}$ m
Micro-toryx eye radius	r_0	$1.408\,970\,164 \times 10^{-15}$ m
Toryx basic frequency	f_0	$3.386\,406\,102 \times 10^{+22}$ s^{-1}
Toryx quantization constant	Λ	137
Rydberg constant	R_x	$1.097\,373\,157 \times 10^{+07}$ m^{-1}
Inverse spacetime fine structure constant	α_s^{-1}	137.036 015 720
Maximum toryx oscillation factor	Q_{qm}	$2.637\,728 \times 10^{+11}$
Maximum toryx oscillation quantum state	q	26.199742

SUMMARY OF GENETIC CODES

1. TORYX

Exhibit 1.3. Toryx basic genetic codes in absolute units.

- The length of one winding of toryx trailing string L_2 is equal to the length of one winding of toryx leading string L_1:

$$L_2 = L_1 = 2\pi r_1 \qquad (1.3\text{-}1)$$

- The difference between the radius of toryx leading string r_1 and the radius of toryx trailing string r_2 is constant and equals to the toryx eye radius r_0:

$$r_0 = r_1 - r_2 = const. \qquad (1.3\text{-}2)$$

- The spiral velocity of toryx trailing string V_2 is constant at each point of its spiral path:

$$V_2 = \sqrt{V_{2t}^2 + V_{2r}^2} = c = const. \qquad (1.3\text{-}3)$$

Exhibit 1.5. Toryx basic genetic codes in relative units.

- The relative length of one winding of toryx trailing string l_2 is equal to the relative length of one winding of toryx leading string l_1:

-

$$l_2 = l_1 \qquad (1.5\text{-}1)$$

- The toryx relative eye radius b_0 is equal to 1:

$$b_0 = b_1 - b_2 = 1 \qquad (1.5\text{-}2)$$

- The relative spiral velocity of toryx trailing string β_2 is equal to 1 at each point of its spiral path:

$$\beta_2 = \sqrt{\beta_{2t}^2 + \beta_{2r}^2} = 1 \qquad (1.5\text{-}3)$$

Exhibit 6.2. Limitations of degrees of freedom of excited toryces.

The relative radius of leading string of real negative toryx b_1 is equal to:		
Lambda toryx	**Harmonic toryx**	**Golden toryx**
$b_1 = z = 2(n\Lambda)^m$ (6.2-1)	$b_1 = 2 + n$ (6.2-2)	$b_1 = 2 + n/\phi$ (6.2-3)

Exhibit 6.3. Limitations of degrees of freedom of oscillated toryces.

The toryx oscillation factor Q_q is equal to:

$$q = 0: \quad Q_0 = 1$$

$$q = 1, 2, 3..: \quad Q_q = 3\left(\frac{\Lambda}{2(q-1)}\right)^{q-1} \qquad (6.3\text{-}1)$$

Exhibit 6.4. Limitation of degrees of freedom of toryx standing wave.

The aspect ratio of toryx standing wave $\lambda_3/2r_4$ is equal to the toryx quantization constant Λ:

$$\frac{\lambda_3}{2r_4} = \Lambda = \text{const.} \qquad (6.4\text{-}5)$$

The frequency of toryx standing wave f_3 is equal to the frequency of toryx trailing string f_2:

$$f_3 = f_2 \qquad (6.4\text{-}6)$$

Exhibit 6.5. Relative radius of toryx string thickness.

The relative radius of toryx string thickness b_s is equal to:

$$b_s = b_u^{-1} = 1/(2Q_{qm} - 1) = 1.895571 \times 10^{-12} \qquad (6.5\text{-}1)$$

Exhibit 7.1. Basic relationships between toryx physical and spacetime parameters.

- The relative toryx charge e_t / e_0 is equal to the toryx vorticity V:

$$\frac{e_t}{e_0} = V \tag{7.1-1}$$

- The relative toryx gravitational mass m_g / m_e is proportional to the absolute value of the toryx vorticity $|V|$:

$$\frac{m_g}{m_e} = Q_q |V| \tag{7.1-2}$$

Exhibit 8.1.1. Toryx Uncertainty Principle.

The product of the **toryx intensity** I and the cycle time of the toryx leading string T_1 must be either equal or smaller than the Planck constant h over 2π as given by the equation:

$$I T_1 \leq \frac{h}{2\pi} = \frac{e_0^2}{4\pi c \varepsilon_0 \alpha_s} \tag{8.1-1}$$

Exhibit 8.1.2. The ultimate maximum relative radius b_{1u} of leading strings of real negative toryx in respect to the maximum toryx oscillation factor Q_{qm}.

$$b_{1u} = Q_{qm} = 2.637728 \times 10^{11} \tag{8.1-4}$$

Exhibit 8.3.1. Tron spacetime quantity absorbed or released by toryces.

The **spacetime quantity** S_t and $\breve{S}_t$ absorbed or released by the real and imaginary toryces are assumed to be proportional to the product of their vorticities V and the square of realities R as given by the equations:

$$S_t = Q_q V R^2 = -\frac{Q_q (b_1 - 1)(2b_1 - 1)}{b_1}$$

$$\breve{S}_t = \breve{Q}_q \breve{V} \breve{R}^2 = -\frac{\breve{Q}_q (\breve{b}_1 - 1)(2\breve{b}_1 - 1)}{\breve{b}_1} \tag{8.3-1}$$

Exhibit 8.3.2. Tron spacetime quantity conservation law.

According to the tron spacetime conservation law, the spacetime quantity S of a tron made up of N real and $\breve{N}$ imaginary positive and negative toryces with the respective toryx spacetime quantities S_t and $\breve{S}_t$ must approach the minimum toryx limits b_u^{-1} as given by the equation:

$$S = (S_t N + \breve{S}_t \breve{N}) \rightarrow \pm b_u^{-1}$$

(8.3-2)

Exhibit 10.5.1. Quantum spacetime levels of the Multiverse.

Spacetime levels	Exponential excitation quantum states m of trons			
	Singulatron	**Ethertron**	**Electron**	**Positron**
L0	$m = 0$	$m = 0$	$m = 1$	$m = 1$
L1	$\boldsymbol{m = 1}$	$\boldsymbol{m = 1}$	$\boldsymbol{m = 2}$	$\boldsymbol{m = 2}$
L2	$m = 2$	$m = 2$	$m = 3$	$m = 3$
L3	$m = 3$	$m = 3$	$m = 4$	$m = 4$
L4	$m = 4$	$m = 4$	$m = 5$	$m = 5$
L5	$m = 5$	$m = 5$	$m = 6$	$m = 6$

Exhibit 12.2.1. Macro-toryx genetic code in absolute units.

- The length of one winding of trailing string L_2 is equal to the length of one winding of leading string L_1:

$$L_2 = L_1 = 2\pi r_1$$

(12.2-1)

- The macro-toryx eye radius r_{jb} is constant:

$$r_{jb} = r_1 - r_2 = const.$$

(12.2-2)

- The spiral velocity of trailing string V_2 is constant at each point of its spiral path:

$$V_2 = \sqrt{V_{2t}^2 + V_{2r}^2} = c = const.$$

(12.2-3)

Exhibit 12.2.2. Macro-toryx genetic codes in relative units.

- The relative length of one winding of trailing string l_2 is equal to the relative length of one winding of leading string l_1:

$$l_2 = l_1 \tag{12.2-4}$$

- The macro-toryx relative eye radius b_{jb} is equal to 1:

$$b_{jb} = b_1 - b_2 = 1 \tag{12.2-5}$$

- The relative spiral velocity of trailing string β_2 is equal to 1 at each point of its spiral path:

$$\beta_2 = \sqrt{\beta_{2t}^2 + \beta_{2r}^2} = 1 \tag{12.2-6}$$

Exhibit 12.5. Spacetime acceleration of a body.

Spacetime acceleration a_{bs} of the body B in respect to a center of the body A is equal to:

$$a_{bs} = \frac{V_{1A}^2 - V_B^2}{r_{1A}} = \frac{V_{1A}^2(1 - \gamma_V^2)}{r_{1A}} \tag{12.5-1}$$

where γ_V is the velocity ratio that is equal to:

$$\gamma_V = V_B / V_1 \tag{12.5-2}$$

Exhibit 12.6. The quantization parameter Z of a macro-toryx.

The quantization parameter Z of a macro-toryx is expressed by the equation:

$$b_1 = Z = 2(n\Lambda)^2 \tag{12.6-1}$$

Exhibit 12.7. The galaxy law of Star Motion.

- The eye radius r_{jg} of a macro-toryx associated with a star is directly-proportional to the galaxy mass m_G contained inside a sphere with the radius equal to the star orbital radius r_s:

$$r_{jg} = \frac{k_g m_G G}{2c^2}$$

(12.7-1)

- The ratio s of the galaxy mass m_G contained inside a sphere with the radius equal to the star orbital radius r_s is constant:

$$\frac{m_G}{r_s} = s = const.$$

(12.7-2)

Where k_g = coefficient related to a specific galaxy.

2. HELYX

Exhibit 13.3. Helyx basic genetic codes in absolute units.

- The frequencies of the first and second levels of helyx trailing string $\widetilde{f}_1$ and $\widetilde{f}_2$ are equal to one another:

$$\widetilde{f}_1 = \widetilde{f}_2 \tag{13.3-1}$$

- The helyx eye radius $\widetilde{r}_0$ is equal to the toryx eye radius r_0 that is constant:

$$\widetilde{r}_0 = r_0 = \widetilde{r}_1 - \widetilde{r}_2 = const. \tag{13.3-2}$$

- The spiral velocity $\widetilde{V}_2$ of the second level helyx trailing string is constant and equals to the velocity of light c at each point of its spiral path:

$$\widetilde{V}_2 = \sqrt{\widetilde{V}_{2t}^2 + \widetilde{V}_{2r}^2} = c = const. \tag{13.3-3}$$

Exhibit 13.5. Helyx basic genetic codes in relative units.

- The relative frequencies of the first and second levels of helyx trailing string $\widetilde{\delta}_1$ and $\widetilde{\delta}_2$ are equal to one another:

$$\widetilde{\delta}_1 = \widetilde{\delta}_2 \tag{13.5-1}$$

- The difference between the relative radius of helyx leading string $\widetilde{b}_1$ and the relative radius of helyx trailing string $\widetilde{b}_2$ is equal to 1:

$$\widetilde{b}_1 - \widetilde{b}_2 = 1 \tag{13.5-2}$$

- The relative spiral velocity $\widetilde{\beta}_2$ of the helyx level 2 is equal to 1 at each point of its spiral path:

$$\widetilde{\beta}_2 = \sqrt{\widetilde{\beta}_{2t}^2 + \widetilde{\beta}_{2r}^2} = 1 \tag{13.5-3}$$

Exhibit 17.1. The spacetime intensity conservation law.

The real helyx spacetime intensity $\widetilde{I}$ is proportional to the relative frequency of real helyx trailing string $\widetilde{\delta}_2$. It is equal to a difference between the real toryx spacetime intensities I_k and I_j of its parental toryx in the higher and lower quantum states k and j:

$$\widetilde{I} = I_k - I_j = 4\alpha_s^{-1}\widetilde{\delta}_2 \tag{17.1-1a}$$

Similarly, the imaginary helyx spacetime intensity $\widetilde{\widetilde{I}}$ is equal to:

$$\widetilde{\widetilde{I}} = \widecheck{I}_k - \widecheck{I}_j = 4\alpha_s^{-1}\widetilde{\widetilde{\delta}}_2 \tag{17.1-1b}$$

REFERENCES

LIST OF PUBLICATIONS CONSULTED

A

Aczel, A.D., *The Mystery of the Aleph*, Washington Square Press, Published by Pocket Books, New York, London, Toronto, Sydney, 2001.

Aczel, A.D., *Entanglement*, A Plume Book, Penguin Group, New York, 2003.

Adair, R.K., *The Great Design – Particles, Fields and Creation*, Oxford University Press, New York, Oxford, 1987.

Agassi, J., *Faraday as a Natural Philosopher*, The University of Chicago Press, Chicago, IL, 1971.

Aiton, E.J., *The Vortex Theory of Planetary Motions*, American Elsevier, Inc., New York, 1972.

Aiton, E.J., *Leibniz - A Biography*, Adam Hilger Ltd, Bristol and Boston, 1985.

Akimov, A.E. and Shipov, G.I., "Torsion Fields and Their Experimental Manifestations," *Proc. of the Int'l Conference on New Ideas in Natural Sciences*, St. Petersburg, Russia, June 1996.

Akimov, A.E. and Tarasenko, V.Y., "Models of Polarized States of the Physical Vacuum and Torsion Fields," *Fizika*, No. 3, March 1992.

Albert, D.Z., "Bohm's Alternative to Quantum Mechanics," *Scientific American*, May 1994.

Albert, D.Z., *Quantum Mechanics and Experience*, Harvard University Press, Cambridge, Massachusetts, London, England, 1992.

Alexander, A., *Infinitesimal – How a Dangerous Mathematical Theory Shaped the Modern World*, Scientific American / Farrar, Straus and Giroux, New York, 2014.

Alexandersson, O., *Living Water - Victor Schauberger and the Secrets of Natural Energy*, Gateway Books, Bath, UK, 1996.

Alfven, H., *Worlds - Antiworlds: Antimatter in Cosmology*, W.H. Freeman and Company, San Francisco, 1966.

Allen, H.S., "The Case for a Ring Electron, *Proc. Phys. Soc. London*, vol. 31, pp 49-68 (1919).

Allen, J., *As a Man Thinketh*, St. Martin's Essentials, New York, 2019.

Allen, R.E., *Greek Philosophy: Thales to Aristotle*, The Free Press, New York, 1966.

Allexander, A., *Infinitesimal*, Scientific American / Farrar, Straus and Giroux, New York, 2014.

Al-Khalili, J., *Quantum – A Guide for the Perplexed*, Weidenfeld & Nicolson Orion House, London, 2004.

Andrade, E.N. da C., *Rutherford and the Nature of the Atom*, Doubleday & Company, Inc., New York, 1964.

Andrews, B., "Outsmarting Einstein," *Discover*, April 2015.

Andrews, B., "Periodic Table," *Discover*, August 2019.

Andrews, B., "Mind-Controlled Rats Are Now a Thing," *Discover*, January/February 2020.

Andrews, B., "A New Tool Turns Up Surprise Slash Across Saturn's Larges Moon," *Discover*, January/February 2020.

Andrulis, E.D., "Theory of the Origin, Evolution, and Nature of Life," (2012), *Life* **2012**, *2*, 1-30.

Andersen, R., The Intention Machine," *Scientific American*, April 2019.

Ariely, D. and Garcia-Rada, X., "Contagious Dishonesty," *Scientific American*, September 2019.

Arp, H., *Seeing Red - Redshifts, Cosmology and Academic Science*, Apeiron, Montreal, Canada, 1998.

Ash, D. and Hewitt, P., *The Vortex - Key to Future Science*, Gateway Books, Bath, England, 1991.

B

Babbitt, E.D., *The Principles of Light and Color*, Babbitt & Co., Kessinger Legacy Reprints, 1878.

Baggott, J., *The Meaning of Quantum Theory*, Oxford University Press, Oxford, UK, 1992.

Baggott, J., *Quantum Space – Loop Quantum Gravity and the Search for the Structure of Space, Time, and the Universe*, Oxford University Press, Oxford, UK, 2018.

Baker, J., *50 Physics Ideas You Really Need to Know,* Quercus, London, 2007.

Barrow, J.D. and Silk, J., *The Left Hand of Creation*, Oxford University Press, New York, 1983.

Barrow, J.D., *The Constants of Nature – The Numbers that Encode the Deepest Secrets of the Universe*, Vintage Books A Division of Random House, Inc., New York, 2002.

Barrow, J.D., *The Infinite Book – A Short Guide to the Boundless, Timeless and Endless*, Pantheon Books, New York, 2005.

Barrow, J.D., *New Theories of Everything – The Quest for Ultimate Explanation*, Oxford University Press Inc., New York, 2007.

Barrow, J.D. and Tipler, F., *The Anthropic Cosmological Principle*, Oxford University Press, Oxford, New York, 2009.

Bartusiak, M., "Loops of Space," *Discover*, April 1993.

Bartusiak, M., "Gravity Wave Sky," *Discover*, July 1993.

Bastrukov, S.I. at al, "Spiral Magneto-Electron Waves in Interstellar Gas," Journal of Experimental and Theoretical Physics, Volume 93, pp 671-676, October 2001.

Beckmann, P., *A History of PI*, Dorset Press, New York, 1989.

Beiser, G., *The Story of Gravity - An Historical Approach to the Study of the Force That Holds the Universe Together*, E.P. Dutton & Co., Inc., New York, 1968.

Bekenstein, J.D., "Information in the Holographic Universe," *Scientific American*, August 2003.

Bender, E., "Global Warning," *Scientific American*, July 2014.

Bentov, I., *Stalking the Wild Pendulum – On the Mechanics of Consciousness*, Density Book, Rochester, New York, 1977.

Bentov, I., *A Brief Tour of Higher Consciousness – A Cosmic Book on the Mechanics of Creation*, Density Book, Rochester, Vermont, 2000.

Bergman, D.L. and Wesley, J.P., "Spinning Charge Ring Model of Electron Yielding Anomalous Magnetic Moment," *Galilean Electrodynamics*, Vol. 1, No. 5, Sept./Oct. 1990.

Birch, B.B., *Power Yoga – The Total Strength and Flexibility Workout,* A Fireside Book. Published by Simon & Schuster, New York, 1995.

Berke, J.P., Author, Editor, *Nanotubes and Nanowires (Selected Topics in Electronics and Systems*, World Scientific Publishing Company, Singapore, 2007.

Bernauer, J.C. and Pohl, R., "The Proton Radius Problem," *Scientific American*, February 2014.

Bertolero, M., "How Matter Becomes Mind," *Scientific American*, July 2019.

Biedermannn, H., *Dictionary of Symbolism –Cultural Icons and the Meaning behind Them*, A Meridian Book, New York, 1994.

Bhadkamkar, A. and Fox, H., "Electron Charge Cluster Sparking in Aqueous Solutions," *Journal of New Energy*, Vol. 1, No. 4, 1996.

Bhaumik, M., *Code Name God – The Spiritual Odyssey of a Man of Science*, The Crossroad Publishing Company, New York, 2005.

Bhaumik, M., *The Cosmic Detective – Exploring the Mysteries of our Universe*, Penguin Books, Ltd., London, England, 2008.

Billings, L., "Center of Gravity," *Scientific American*, November 2019.

Blackwood, O.H., et al, *An Outline of Atomic Physics*, John Wiley & Sons, Inc., New York, 1955.

Blanco, D.B., "A melting Planet," *Discover*, January/February 2020.

Bloyd, J.G., *Broken Arrow of Time - Rethinking the Revolution in Modern Physics*, Writers Club Press, San Jose, CA, 2001.

Born, M., *Atomic Physics*, Dover Publications, Inc., Mineola, NY, 1989.

Borchardt, G., *Infinity Universe Theory*, Progressive Science Institute, Berkley, CA, 2018.

Boscovich, R.J., *A Theory of Natural Philosophy*, The M.I.T. Press, Cambridge, MA, 1966.

Boslough, J., *Stephen Hawking's Universe - An Introduction to the Most Remarkable Scientist of Our Time*, Quill/William Morrow, New York, 1985.

Boslough, J., *Masters of Time - Cosmology at the End of Innocence*, Addison-Wesley Publishing Company, Reading, MA, 1992.

Bostick, W., "Mass, Charge, and Current: The Essence of Morphology," *Physics Essays*, Vol. 4, No. 1, pp. 45-59, March 1991.

Bowers, B., *Michael Faraday and Electricity*, Priory Press Ltd., London, 1974.

Bradford, M., *The Healing Energy of Your Hands,* The Crossing Press, Inc,, Freedom, California, 1995.

Brennan, R.P., *Heisenberg Probably Slept Here – The Lives, Times and Ideas of the Great Physicists of the 20th Century*, John Wiley & Sons, Inc., New York, 1997.

Broglie, L., de, *The Revolution in Physics*, The Noonday Press, New York, 1953.

Broglie, L., de, *New Perspectives in Physics*, Basic Books, Inc. Publishers, New York, 1962.

Burger, T.J., *Nature* **271**, 402, 1978.

Burrell, "T., "Mini-Brains Make New Waves," *Discover*, January/February 2020.

C

Cale, D.L., *The Simplest Possible Universe*. Matheia Society Press, Uniontown PA, 2017.

Calladine, C.R. and Drew, H.R., *Understanding DNA – The Molecule and How It Works*, Second Edition, Academic Press, New York, 2002.

Cambier, J-L., at al, "Theoretical Analysis of the Electron Spiral Toroidal Concept," NASA/CR-2000-210654, Dec. 2000.

Capra, F., *The Web of Life*, Anchor Books/Random House, Inc., 1997.

Capra, F., *The Tao of Physics*, Shambhala Publications, Inc., 1999.

Carter, J., *The Other Theory of Physics - A Non-Field Unified Theory of Matter and Motion*, Absolute Motion Press, 2000.

Carrigan, Jr., R.A. and Tower, W.P., *Particles and Forces at the Heart of the Matter,* W.H. Freeman and Company, New York, 1990.

Carroll, R.L., *The Energy of Physical Creation,* The Carroll Research Institute, P.O. Box 3425, Columbia, S.C., 29230, 1985.

Carter, J., *The Other Theory of Physics – A Non-Field Unified Theory of Matter and Motion,* Absolute Motion Press, Enumclaw, Washington, 2000.

Cartledge, P., *Democritus*, Routledge, New York, 1999.

Cassel, G.H, Billig, M.D., & Randall, H.G., *The Eye Book – A Complete Guide to Eye Disorders and Health,* A John Hopkins Press Health Book, 1998.

Cassel, G.H., *The Eye Book – A Complete Guide to Eye Disorders and Health,* Second edition, A John Hopkins Press Health Book, 2021.

Caspar, M., *Kepler*, Abelard-Schuman, London and New York, 1992.

Cecil, T.E. and Chern, S., *Tight and Taut Submanifolds*, Cambridge University Press, New York, 1997.

Chalidze, V., *Mass and Electric Charge in the Vortex Theory of Matter*, Universal Publishers, 2001.

Chen, C., at al, "Equilibrium and Stability Properties of Self-Organized Electron Spiral Toroid," Physics of Plasma, Volume 8, Number 10, October 2001.

Chen, Y., Editor, *Nanotubes and Nanosheets: Functionalization and Applications of Boron Nitride and Other Nanomaterials,* CRC Press, London, 2015.

Chia, M., *Chia Self-Massage – The Taoist Way of Rejuvenation,* Healing Tao Books, Huntington, New York, 1986.

Ching, T.T., *The Book of Balance,* The University Science and Philosophy, www.philosophy.org/www.twilightclub.org, 2002.

Chu, T., *Human Purpose and Transhuman Potential – A Cosmic Vision for Future Evolution*, Origin Press, San Rafael, CA, 2014.

Chuen. L.K., *Step – By – Step Tai Chi.* A Fireside Book. Published by Simon & Shuster, Inc., New York, 1994.

Citro, M., *The Basic Code of the Universe – The Science of the Invisible in Physics, Medicine, and Spirituality,* Park Street Press, Rochester, Vermont, 2011.

Clark, G., *The Man Who Tapped the Secrets of the Universe,* The University of Science and Philosophy, Swannanoa, Waynesboro, 2000.

Clawson, C.C., *Mathematical Sorcery – Revealing the Secrets of Numbers*, Perseus Books, Cambridge, MA, 2001.

Clawson, C.C., *Mathematical Mysteries – The Beauty and Magic of Numbers*, Perseus Publishing, Cambridge, MA, 1999.

Clegg, B., *Before the Bing Bang*, St. Martin's Griffin, New York, 2009.

Close, F., *Neutrino,* Oxford University Press, 2010.

Close, F., Marten, M., & Sutton, C., *The Particle Explosion*, Oxford University Press, New York, 1994.

Close, J.P, "Ancient Plagues Shaped the World," *Scientific American*, November 2020.

Coats, C., *Living Energies*, Gateway Books, Bath, UK, 1996.

CODATA Recommended Values, *The NIST Reference on Constants, Units and Uncertainties*, 2011.

Coe, D., "Back in Time," *Scientific American*, November 2018.

Collins, H. and Pinch T., *The Golem – What Everyone Should Know about Science,* Cambridge University Press, 1993.

Collins, F.S., "The First Gene-Edited Babies Turn 1," *Discover*, January/February 2020.

Compton, A., "The Size and Shape of the Electron," *Phys. Rev. Second Series*, vol. 14, no.3, pp 247-259, (1919).

Conniff, R., "The Last Resort," *Scientific American*, January 2019.

Consa, O., "Helical Model of the Electron," The General Science Journal, June 2014.

Cook, N., *The Hunt for Zero Point - Inside the Classified World of Antigravity Technology*, Broadway Books, New York, 2001.

Cook, T.A., *The Curves of Life*, Dover Publications, Inc., New York, 1979.

Coxeter, H.S.M., *Introduction to Geometry*, John Wiley & Sons, Inc., New York, 1961.

Coxeter, H.S.M., *The Beauty of Geometry*, Dover Publications, Inc., New York, 1968.

Coyne, G. and Heller, M, *A Comprehensible Universe*, 2008.

Crandall, B.C., *Nanotechnology – Molecular Speculations on Global Abundance,* The MIT Press, Cambridge, Massachusetts, London, England, 1996.

Crew, H., *The Wave Theory of Light - Memoirs by Huygens, Young and Fresnel*, American Book Company, New York, 1900.

Cubitt, T.S., Perez-Garcia, D. and Wolf, M., "The Un(solv)able Problem," *Scientific American*, October 2018.

Cushing, J.T., *Philosophical Concepts in Physics – The History Relations between Philosophy and Scientific Theories,* Cambridge University Press, 1998.

D

Dahlman, J., "All the World's Data Could Fit in an Egg," *Scientific American*, June 2019.

Dalton, J., et al, *Foundations of the Atomic Theory: Comprising Papers and Extracts*, Alembic Club, Edinburgh, UK, 1968.

Danaylov, N., *Conversations with the Future – 21 Visions for the 21st Century*, Singularity Media, 2016.

D'Auria, S., *Introduction to Nuclear and Particle Physics*, Springer Nature, Switzerland AG, 2018.

Davies, P., *About Time - Einstein's Unfinished Revolution*, Simon & Schuster, New York, 1995.

Davies, P., *The Goldilocks Enigma – Why is the Universe Just Right for Life?* A Mariner Book, Houghton Mifflin Company, New York, 2008.

Davies, P.C.W. and Brown, J., *Superstrings - A Theory of Everything?* Cambridge University Press, Cambridge, UK, 1988.

Davies, P. and Gribbin, J., *The Matter Myth - Dramatic Discoveries That Challenge Our Understanding of Physical Reality*, Touchstone Book/Simon & Schuster, New York, 1992.

Day, W., *Bridge from Nowhere - The Photonic Origin of Matter*, Rhombics, Cambridge, MA, 1996.

Day, W., *A New Physics - Foundation for New Directions*, Cambridge, MA, 2000.

Dengler, R., "A million Species in Danger," *Discover*, January/February 2020.

Dengler, R., "Brains Brought Back to Life," *Discover*, January/February 2020.

Derbyshire, J., *Unkn( )wn Quantity - A Real and Imaginary History of Algebra*, A Plume Book, Published by Penguin Group, New York, 2007.

Deshpandle, A. and Yoshida, R., "The Deepest Recesses of the Atom," *Scientific American*, June 2019.

Di Mario, D., "Electrogravity: A Basic Link Between Electricity and Gravity," *Speculations in Science and Technology*, Vol. 20, No. 4, Dec. 1997.

Dibner, B., *Oersted - And the Discovery of Electromagnetism*, Blaisdell Publishing Company, New York, 1962.

Dijksterhuis, E.J., *Archimedes*, Princeton University Press, Princeton, N.J., 1987.

Dirac, PAM, "Quantized Singularities in the Electromagnetic Fields," Scribd.com. 1931-05-29.

Dixon, C., "Monster Waves," *Scientific American*, August 2018.

Dixon, R., *Mathographics*, Dover Publications, Inc, New York, 1991.

Dmitriyev, V.P., "Mechanical Analogy for the Wave - Particle: Helix on Vortex Filament," *Apeiron*, Vol. 8, No. 2, April 2001.

Domb, C., *Clerk Maxwell and Modern Science - Six Commemorative Lectures*, The Athlone Press, University of London, UK, 1963.

Dowdye, E.H., *Discourses & Mathematical Illustrations Pertaining to the Extinction Shift Principle Under Electrodynamics of Galilean Transformations,* Second Edition, Washington DC, 2001.

Drake, S., *Galileo at Work, His Scientific Biography*, The University of Chicago Press, Chicago, 1978.

Dresselhaus, M.S. and Eklund, P.C., *Science of Fullerenes and Carbon Nanotubes: Their Properties and Applications*, Academic Press, 1996.

Drew, H.R., "The Electron as a Four-Dimensional Helix of Spin-1/2 Symmetry," *Physics Essays*, Vol. 12, No. 4, 1999.

Driscoll, R.B., *United Theory of Ether, Field and Matter*, Published by Author, Portland, OR, 1964.

Driscoll, R.B., *United Theory of Ether, Field and Matter (supplement)*, Published by Author, Oakland, CA, 1965.

Duchin, M., "Geometry V. Gerrymandering," *Scientific American*, November 2018.

Duffy, M.C. and Levy, J., Editors, Krasnoholovets, V., Executive Editor, *Ether Space-Time & Cosmology, Volume 1 – Modern Concepts, Relativity and Geometry*, PD Publications, Liverpool, UK, 2008.

Duffy, M.C. and Levy, J., Editors, *Ether Space-Time & Cosmology, Volume 2 – New Insights into a Key Physical Medium*, Aperon, Montreal, Canada, Published by C. Roy Keys, Inc, 2009.

Duffy, M.C. and Levy, J., Editors, *Ether Space-Time & Cosmology, Volume 3 – Physical Vacuum, Relativity and Quantum Physics*, Aperon, Montreal, Canada, Published by C. Roy Keys, Inc, 2009.

Duncan, J.C., *Astronomy – A Textbook*, Fifth Edition, Harper & Brothers Publishers, New York, 1926.

Dyer, W.W., *Your Erroneous Zones*, Harper, New York, 2001.

Dyson, M.J. and Cousins, L., *Welcome to the Moon,* Aldrin Family Foundation, Melbourne, FL, 2019.

E

Eagleton. T., *The Meaning of Life – A Very Short Introduction*, Oxford University Press, 2007.

Eckhart, L., *Four-Dimensional Space*, Indiana University Press, Bloomington, 1968.

Edwards, E.B., *Pattern and Design with Dynamic Symmetry*, Dover Publications, Inc., New York, 1932.

Edwards, L., *The Vortex of Life – Nature's Patterns in Space and Time*, Floris Books, Edinburgh, UK, 2006.

Ehrlich, R., *Crazy Ideas in Science - If You Might Even be True,* Princeton University Press, Princeton and Oxford, 2002.

Einstein, A. and Hopf, L., Ann. Phys., 33, 1096 (1910a): Ann. Phys., 33, 1105, 1910b.

Einstein, A., *Out of My Later Years*, A Citadel Press Book-Carol Publishing Group, New York, NY, 1991.

Einstein, A., *Relativity - The Special and the General Theory*, Crown Publishers, Inc., New York, 1961.

Einstein, A., Infeld, L., *The Evolution of Physics - From Early Concepts to Relativity and Quanta*, A Touchstone Book/Simon & Schuster, New York, 1966.

Einstein, A., "Aether and the Theory of Relativity," (Address on May 5, 1920, at the University of Leyden), *Journal of New Energy*, Vol. 7, No. 1, 2003.

Elgin, D., *The Living Cosmos*, ReVision, The Journal of Consciousness and Change, Volume 11, Number 1, Summer 1988.

Elgin, D., *The Living Universe – Where are We? Who are We? Where are We Going?* Berrett-Koehler Publishers, Inc., San Francisco, CA, 2009.

Epstein, L.C., *Thinking Physics Is Gedanken Physics*, Insight Press, San Francisco, CA, 1983.

Epstein, L.C., *Relativity Visualized*, Insight Press, San Francisco, CA, 1992.

F

Fairlay, P., "The H_2 Solution," *Scientific American*, February 2020.

Farndon, J,. *The Great Scientists – From Euclid to Stephen Hawking,* Metro Books, New York, 2007.

Farrington, B., *Greek Science - Its Meaning for Us*, Penguin Books, Baltimore, MD, 1971.

Feber, A., "Supertwistors and Conformal Supersymmetry," *Nuclear Physics B* **132**: 55-64, 1978.

Feldenkrais, M., *Awareness Through Movement,* HarperCollins Publishers, NY, 1990.

Ferguson, K., *Stephen Hawking - Quest for a Theory of Everything*, Bantam Books, New York, 1992.

Feynman, R.P., *Six Easy Pieces and Six Not-So-Easy Pieces,* Perseus Publishing, Cambridge, Massachusetts, 1995.

Finkbeiner, D., Su, Meng and Malyshev. D., "Giant Bubbles of the Milky Way," *Scientific American*, July 2014.

Finkbeiner, A., "Messengers from the Sky," *Scientific American*, September 2019.

Finkbeiner, A., "Orbital Aggression," *Scientific American*, November 2020.

Fischhoff, B., "Tough Calls," *Scientific American*, May 2018.

Flander, T.V., *Dark Matter, Missing Planets & New Comets – Paradoxes Resolved, Origins Illuminated*, Revised Edition, North Atlantic Books, Berkley, California, 1993.

Flood, R. and Lockwood, M., *The Nature of Time*, Basil Blackwell, Inc., Cambridge, MA, 1990.

Folger, T., "Tangled Up In Strings – Two Books Say That Today's Theoretical Physicists Are Way Off Course," *Discover*, September 2006.

Folger, T., "Your Daily Dose of Quantum," *Discover*, November 2018.

Folger, T., "Crossing the Quantum Divide," *Scientific American*, July 2018.

Folger, T., "Quantum Gravity in the Lab," *Scientific American*, April 2019.

Ford, K.W., *101 Quantum Questions*, Harvard University Press, Cambridge, Massachusetts, 2011.

Fortenberry, R., First Molecule in the Universe," *Scientific American*, February 2020.

Fowler, P.W. and Manolopoulos, D.E., *An Atlas of Fullerenes*, Dover Publications, 2007.

Frank, P., *Einstein - His Life and Times*, Da Capo Press, Inc., New York, 1947.

Frankl, V.E., *Man's Search for Meaning*, Beacon Press, Boston, 2006.

Fraser, et al, *The Search for Infinity*, Facts on File, Inc., New York, 1995.

Freedman, D.H., "The Mysterious Middle of the Milky Way," *Discover*, November 1998.

Freeman, K. and McNamara, G., *In Search of Dark Matter*, Springer Praxis Publishing, Chichester, UK, 2006.

Friedman, N., *Bridging Science and Spirit - Common Elements in David Bohm's Physics, The Perennial Philosophy and Seth*, Living Lake Books, St. Louis, MO, 1994.

Fritzsch, H., *Quarks - The Stuff of Matter*, Basic Books, Inc., New York, 1983.

Fritzsch, H., *The Creation of Matter - The Universe From Beginning to End*, Basic Books, Inc., New York, 1984.

Funk & Wagnalls New Encyclopedia, Funk & Wagnalls, Inc., USA, 1966.

G

Gabrieli, J., "A Look Within," *Scientific American*, March 2018.

Gamow, G., *The Great Physicists from Galileo to Einstein,* Dover Publications, Inc., New York, 1988.

Gamow, G., *Thirty Years That Shook Physics - The Story of Quantum Theory*, Dover Publications, Inc., New York, 1985.

Gamow, G., *One, Two, Three ... Infinity - Facts and Speculations of Science*, Dover Publications, Inc., New York, 1988.

Gardner, M., *New Mathematical Diversions from Scientific American*, Simon and Schuster, New York, 1966.

Gardner, M., *Knotted Doughnuts and Other Mathematical Entertainments*, W.H. Freeman and Company, New York, 1986.

Gasperini, M., *The Universe Before the Big Bang,* Springer, Berlin, 2010.

Gauthier, R., "Faster-than-light quantum models of the photon and the electron", in M. S. El-Genk, *(*ed.) *"Space Technology and Applications International Forum – STAIF 2007"*, American Institute of Physics 978-0-7354-0386-4/07, p1099-1108, 2007.

Gauthier, R., "Transluminal energy quantum models of the photon and the electron", in R.L. Amoroso, P. Rowlands & L.H. Kauffman (eds.) *The Physics of Reality: Space, Time, Matter, Cosmos, 8th Symposium in Honor of Mathematical Physicist Jean-Pierre Vigier*, Hackensack: World Scientific, 2013.

Gauthier, R., "A transluminal energy quantum model of the cosmic quantum", in R.L. Amoroso, P. Rowlands & L.H. Kauffman (eds.) *The Physics of Reality: Space, Time, Matter, Cosmos, 8th Symposium in Honor of Mathematical Physicist Jean-Pierre Vigier*, Hackensack: World Scientific, 2013.

Gautreau, R. and Savin, W., *Schaum's Outline of Theory and Problems of Modern Physics,* McGeaw-Hill, New York, 1978.

Gazale, M.J., *Gnomon*, Princeton University Press, Princeton, N.J., 1999.

Geerlings, G.K., *Wrought Iron In Architecture – An Illustrated Survey,* Dover Publications, Inc, New York, 1983.

Gell-Mann, M., *Quark and the Jaguar – Adventures in the Simple and the Complex*, A W, H. Freeman/Owl Book, Henry Holt and Company, LLC, New York, 1994.

Gell-Mann, M., *Complexity*, Vol. 1, no 5, John Wiley and Sons, Inc., New York, 1995/96.

Genz, H., *Nothingness - The Science of Empty Space*, Perseus Books, Reading, MA, 1999.

Gerber, R.G., *Vibrational Medicine – The #1 Handbook of Subtle-Energy Therapies Medicine,* Third Edition, Bear & Company, Rochester, Vermont, 2001.

Geymonat, L., *Galileo Galilei: A Biography and Inquiry into His Philosophy of Science*, McGraw-Hill

Book Company, 1965.

Ghosh, A., *Origin of Inertia*, Apeiron, Montreal, Canada, 2000.

Ghyka, M., *The Geometry of Art and Life*, Dover Publications, Inc, New York, 1977.

Giddlings, S.B., "Escape from Black Hole," *Scientific American*, December 2019.

Gillispie, C.C., *Dictionary of Scientific Biography*, Vol. IV, Charles Scribner's Sons, New York, 1971.

Ginzburg, V.L., *Theoretical Physics and Astrophysics*, Pergamon Press, 1979.

Ginzburg, V.L., *Physics and Astrophysics. A Selection of Key Problems*, Pergamon Press, 1985.

Gleick, J., *Chaos - Making a New Science*, Penguin Books, New York, 1988.

Gleick, J., *Genius - The Life and Science of Richard Feynman*, Pantheon Books, New York, 1992.

Gould, R.R., *Universe in Creation – A New Understanding of the Big Bang and the Emergence of Life*, Harvard University Press, Cambridge Massachusetts, London, England, 2018.

Gorini, C.A., *Geometry*, Facts on File, Inc., New York, 2003.

Gordon, R., *Your Healing Hands – The Polarity Experience,* North Atlantic Books, Berkeley CA, 2004.

Gordon, R., *Quantum Touch – The Power to Heal,* North Atlantic Books, Berkeley CA, 2006.

Goswami, A., *The Self-Aware Universe - How Consciousness Creates the Material World*, Penguin Putnam Inc., 1993.

Graver, J.E., "The Structure of Fullerene Signatures," *DIMACS Series in Discrete Mathematics and Theoretical Computer Science*, American Mathematical Society, 2005.

Gray, A., *Lord Kelvin - An Account of His Scientific Life and Work*, E.P. Dutton & Co., 1908.

Gray, A., *Modern Differential Geometry of Curves and Surfaces*, CRC Press, Boca Raton, 1993.

Grego, L., "Broken Shield," *Scientific American*, June 2019.

Greene, B., *The Elegant Universe - Superstrings, Hidden Dimensions, and the Quest for the Ultimate Theory*, W.W. Norton & Company, New York, 1999.

Greene, B., *The Fabric of the Cosmos*, Alfred A. Knopf, New York, 2004.

Gribbin, J., *Q Is For Quantum, An Encyclopedia of Particle Physics*, A Touchstone Book/ Simon & Schuster, New York, 2000.

Gribbin, J., *Schrödinger's Kittens and the Search for Reality,* Little, Brown & Company (Canada) Limited, 1995.

Groves, A., "Repairing the Future," *Discover*, May 2019.

Guillen, M., *Five Equations that Changed the World*, Hyperion, New York, 1995.

Gutsche, J., *Better and Faster – The Proven Path to Unstoppable Ideas*, Currency, New York, 2015.

Gutsche, J., *Exploiting Chaos – 150 Ways to Spark Innovation during Times of Change*, Gotham Books, New York, 2009.

H

Hadhazy, A., "Putting the Relativity to the Test," *Discover*, April 2015.

Hadhazy, A., "The Dark Matter Derby," *Discover*, December 2019.

Hadhazy, A., "10 Experiments That Changed Everything," *Discover*, November 2019.

Haines, K., "The Moon," *Discover*, July/August 2019.

Haines, K., "The Bumpy Road to Launching a Dragon." *Discover*, January/February 2020.

Haisch, B., *The Purpose-Guided Universe – Believing in Einstein, Darwin, and God*, The Career Press, Inc., Franklin Lakales, NJ, 2010.

Haisch, B., *The God Theory - Universes, Zero-Point Fields, and What's Behind It All*, Weser Books, San Francisco, CA, 2006.

Halderman, J.A. and Schwartz, J., "How to Defraud Democracy," *Scientific American*, September 2019.

Hall, A.R., *Isaac Newton - Adventurer in Thought*, Blackwell Publishers, Oxford, UK, 1994.

Hall, S., "Hidden Inferno," *Scientific American*, December 2018.

Hambidge, J., *Practical Applications of Dynamic Symmetry*, The Devin-Adair Company, New York,

1967.

Hanson, R. and Shalm, K., "Spooky Action," *Scientific American*, December 2018.

Harari, Y.N., *Sapiens – A Brief History of Humankind*, Harper Perennial, New York, 2015.

Harari, Y.N., *21 Lessons for the 21ˢᵗ Century*, Spiegel & Grau, New York, 2018.

Hargittai, I., Pickover, C.A., Editors, *Spiral Symmetry*, World Scientific Publishing Co., Singapore, 1992.

Harrington, P.S, *Star Watch*, John, Wiley & Sons, Inc., Hoboken New Jersey, 2003.

Harris, P.J.F., *Carbon Nanotube Science: Synthesis, Properties and Applications,* Cambridge University Press, Cambridge, UK, 2011.

Harrison, L.P., *Meteorology*, National Aeronautics Council, Inc., New York, 1942.

Hatch, E., *Modern Physics from a Classical Scale Perspective- Part 1 Concept Confirmed*, RWWAA Publication, Auburn, California, 2004.

Hawking, S., *Black Holes and Baby Universes and Other Essays*, Bantam Books, New York, 1993.

Hawking, S., *A Brief History of Time - From the Big Bang to Black Holes*, Bantam Books, New York, 1990.

Hawking, S., *On the Shoulders of Giants,* Running Press, Philadelphia, London, 2004.

Hawking, S. and Penrose, R., *The Nature of Space and Time*, Princeton University Press, Princeton, NJ, 2000.

Hay, L.L., *Heal Your Body – The Mental Causes for Physical Illness and Metaphysical Way to Overcome Them,* Hay House, Inc., Carlsbad, California, 1988.

Haynes, K., "Humanity's First Look at a Black Hole," *Discover*, January/February 2020.

Haynes, K., "Race for the Moon," *Discover*, January/February 2020.

Haynes, K., "InSights of Frustrating First Year on Mars," *Discover*, January/February 2020.

Heath, J.L., *The Works of Archimedes*, Dover Publications, Inc., New York, 1953.

Heilbron, J.L., *The Dilemmas of an Upright Man - Max Planck as Spokesman for German Science*, University of California Press, Berkeley, 1986.

Heisenberg, E., *Inner Exile - Recollections of a Life with Werner Heisenberg*, Birkhauser Boston, MA, 1980.

Helmholtz, H., *On the Sensation of Tone – As a Physiological Basis for the Theory of Music,* Dover Publications, New York, 1954.

Henderson, L.D., *The Fourth Dimension and Non-Euclidean Geometry in Modern Art*, Princeton, 1983.

Hershberger, S.., "The Pandemic We Forgot," *Scientific American*, November 2020.

Herzberg, G., *Atomic Spectra and Atomic Structure*, Dover Publications, Inc., New York, 1944.

Hey, N., *Solar System*, Weidenfeld & Nicolson, The Orion Publishing Group, Wellington House, London, UK, 2002.

Hey, T. and Walters, P., *The New Quantum Universe*, Cambridge University Press, UK, 2003.

Hippel von, F., *Citizen Scientist*, A Touchstone Book/Simon & Schuster, New York, 1991.

Ho, A.Y.Q., "Explosion at the Edge," *Scientific American*, December 2020.

Hoffmann, B., *Albert Einstein - Creator and Rebel*, New American Library, New York, 1972.

Hoffman, R.N., "Controlling Hurricanes – Can Hurricanes and Other Severe Tropical Storms be Moderated or Deflected?" *Scientific American*, October 2004.

Hogan, R.C., *Your Eternal Self*, Greater Reality Publications, 2008.

Hogg, M.A., "Radical Change," *Scientific American*, September 2019.

Hölderfin, F., *Human Purpose & the Universal Pursuit of Ecstasy*, Granmore Publications, Exeter, England, 2019.

Hooft, G., *In Search of the Ultimate Building Blocks*, Cambridge University Press, UK, 1997.

Horgan, J., "Gravity Quantized? - A Radical Theory of Gravity Weaves Space from Tiny Loops," *Scientific American*, September 1992.

Hossenfelder, S. and McGaugh, S.S., Is Dark Matter Real?" *Scientific American*, August 2018.

Hotson, D.L., "Dirac's Equation and the Sea of Negative Energy, Part 1," *Infinite Energy*, Vol. 8, Issue 43, 2002.

Hotson, D.L., "Dirac's Equation and the Sea of Negative Energy, Part 2," *Infinite Energy*, Vol. 8, Issue 44, 2002.

Houston-Edwards, K., "Number Games," *Scientific American*, September 2019.

Hullman, J., "Confronting Unknowns," *Scientific American*, September 2019.

Hurley, W.M., *Prehistoric Cordage – Aldine Manuals on Archeology 3,* Taraxacum, Washington, 1979.

I

Icke, V., "From Expansion to Intelligence in the Universe," *Speculations in Science and Technology*, Vol. 14, No. 4, 1991.

Ipsen, D.C., *Archimedes: Greatest Scientist of the Ancient World*, Enslow Publishers, Inc., Hillside, New Jersey, 1988.

Iyengar, B., K., S., *Light Yoga,* Schocken Books. New York, 1966.

J

Jammer, M., *Concepts of Mass in Classical and Modern Physics*, Dover Publications, Inc., Mineola, New York, 1997.

Jammer, M., *Concepts of Force*, Dover Publications, Inc., Mineola, New York, 1999.

Jean, Sir, J, *Science & Music,* Dover Publications, Inc., New York, 1968

Johnson, G., "The Inelegant Universe – Two New Books Argue That It Is Time for String Theory To Give Way," *Scientific American*, September, 2006.

Jefimenko, O.D., *Gravitation and Cogravitation*, Electret Scientific Company, Star City, West Virginia, 2006.

Jones, A.Z. and Robbins, D., *String Theory for Dummies,* John Wiley & Sons, Inc., 2010.

Jones, B.Z., *The Golden Age of Science,* Simon and Schuster, New York, 1966.

Jonsson, I., *Emanuel Swedenborg*, Twayne Publishers Inc., New York, 1971.

Jorgenson, A., "Sex in the Cosmic City," *Discover*, May 2019.

K

Kafatos, M. and Nadeau, R., *The Conscious Universe - Part and Whole in Modern Physical Theory*, Springer-Verlag New York, Inc., New York, 1990.

Kaku, M., *Physics of the Impossible,* Anchor Book, A Division of Random House, Inc., 2008.

Kaku, M., *Visions – How Science Will Revolutionize the 21st Century,* Anchor Books, Doubleday, New York, London, 1997.

Kaku, M., *Beyond Einstein - The Cosmic Quest for Theory of the Universe*, Anchor Books/Doubleday, New York, 1995.

Kaku, M., *Hyperspace*, Anchor Books/Doubleday, New York, 1994.

Kaku, M., *Introduction to Superstrings*, Springer-Verlag, New York, 1988.

Kanarev, F.M., "Model of the Electron," APERON, Vol. 7. Nr. 3-4, July-October, 2000.

Kanarev, F.M., *The Foundation of Physchemistry of Microworld*, Kuban State Agrarian University (KSAU), Krasnodar, Russia, 2002.

Kane, G., *The Particle Garden - Our Universe as Understood by Particle Physicists*, Addison-Wesley Publishing Company, Reading, MA, 1995.

Kanigel, R., *The Man Who Knew Infinity - A life of the Genius Ramanujan,* Washington Square Press, Published by Pocket Books, New York, London, 1991.

Kaplan, R., *The Nothing That Is - A Natural History of Zero*, Oxford University Press, Oxford, New

York, 1999.

Kaplan, R. and Kaplan, E., *The Art of The Infinite*, Oxford University Press, Oxford, New York, 2003.

Kaufmann, W.J., *Black Holes and Warped Spacetime*, W.H. Freeman and Company, San Francisco, CA.

Keats. J, "Nature's Jump Drive." *Discover*, November 2019.

Kenrick, D.T. et al, "The Science of Anti-Science Thinking," *Scientific American*, July 2018.

Kimura, Y.G., *The Book of Balance*, (Translation), The University of Science and Philosophy, Contact Printing, North Vancouver, B.C., Canada, 2002.

Kimura, Y.G., "The Transcendent Unity of Science and Spirituality," *VIA – Vision in Action*, Vol. 2, No. 1 & 2, 2004.

King, B.J., "Deception in the Wild," *Scientific American*, September 2019.

King, M.B., "Vortex Filaments, Torsional Fields and the Zero-Point Energy," *Journal of New Energy*, Vol. 3, No. 2/3, 1998.

King, M.B., "Dual Vortex Forms: The Key to a Large Zero-Point Energy Coherence," *Journal of New Energy*, Vol. 5, No. 2, 2000.

King, M.B., *Quest for Zero Point Energy*, Adventures Unlimited Press, Kempton, IL, 2001.

King, M.B., *Tapping the Zero-Point Energy,* Paraclete Publishing, Provo, Utah, 1989.

Khalsa, S.K., *K.I.S.S – Guide to Yoga,* A Dorling Kindersley Book, London, New York, 2001.

Klemke, E.D. and Cahn, S.M., *The Meaning of Life*, Oxford University Press, 2018.

Klesman, A., "Complex Chemistry of Titan," *Discover*, July/August 2019.

Knight, D.C., *The Science Book of Meteorology*, Franklin Watts, Inc., New York, 1964.

Koch, C., Proust among the Machines," *Scientific American*, December 2019.

Kondrot, K.C., *10 Essentials to Save Your Sight,* Advantage, Charleston. South Carolina, 2012.

Kipnis, J., "The Seven Sense," *Scientific American*, August 2018.

Krauss, L.M., *Quintessence - The Mystery of Missing Mass in the Universe*, Basic Books, New York, NY, 2000.

Krauss, L.M., *A Universe from Nothing*, Atria Paperback, New York, NY, 2012.

Kumar, S., "A Spiral Structure for Elementary Particles," Int. J. Res. Vol. 1, Issue 6, July 2014.

Kumar, S., "Journey of the Universe from Birth to Rebirth with Insight into Unified Interaction of Elementary Particles with Spiral Structure," Int. J. Res. Vol. 1, Issue 9, October 2014.

Kumar, S., "Quantum Spiral Theory," Int. J. Res. Vol. 2, Issue 1, January 2015.

Kumar, S., "Spiral Hashed Information Vessel," International Journal of Scientific & Engineering Research, Vol. 4, Issue 6, April 2015.

Kumar, S., "Spiral Structure of Elementary Particles Analogous to Sea Shells: A Mathematical Description," International Journal of Current Research, Vol. 7, Issue 02, Feb. 2015, p. 12814.

Kumar, S., "Mass-Energy Equivalence in Spiral Structure for Elementary Particles and Balance of Potentials," International Journal of Scientific & Engineering Research. Vol. 6, Issue 6, July 2015.

Kwon, D., "Self-Taught Robots," *Scientific American*, March 2018.

L

Lamb, G.L., "Solutions and the Motion of Helical Curves," *Physical Review Letters*, Vol. 37, No. 5, August 1976.

Lakhtakia, A. and Weiglhofer, W.S., "Time-Dependent Beltrami Fields in Free Space: Dyadic Green Functions and Radiation Potentials," *Physical Review E*, Vol. 49, Number 6, June 1994.

Lakhtakia, A. and Weiglhofer, W.S., "Covariances and Invariances of the Beltrami-Maxwell Postulates," *IEE Proc. - Sci. Meas. Technol.*, Vol. 142, No. 3, May 1995.

Lang, T.G., "Proposed Unified Field Theory – Part II: Protons, Neutrons and Fields," *Galilean Electrodynamics*, Vol. 12, No. 6, Nov./Dec. 2001.

Larsen, R., et al, *Emanuel Swedenborg - A Continuing Vision*, Swedenborg Foundation, Inc., New York,

1988.

Laugwitz, D., *Differential and Riemannian Geometry*, Academic Press, New York, 1965.

Lauwerier, H., *Fractals - Endlessly Repeated Geometrical Figures*, Princeton University Press, Princeton, New Jersey, 1991.

Lederman, L.M. and Teresi, D., *The God Particle*, Bantam Doubleday Dell Publishing Group, Inc., New York, 1993.

Lederman, L.M. and Hill, C.T., *Symmetry and the Beautiful Universe*, Prometheus Books, New York, 2004.

Lederman, L.M. and Hill, C.T., *Quantum Physics for Poets*, Prometheus Books, New York, 2011.

Lederman, L. and Hill, C., *Beyond the God Particle*, Prometheus Books, New York, 2013.

Lerner, E.J., *The Big Bang Never Happened*, Vintage Books, Random House, Inc., New York, 1992.

Lewis, H., *Geometry – A Contemporary Course,* Third Edition, McCormick-Mathers Publishing Company, Cincinnati, Ohio, 1973.

Lewis, J.R., *Scientology*, Cary, NC, Oxford University Press, 2009.

Lewis, T, "COVID-19 Misinformation That Won't Go Away," *Scientific American*, November 2020.

Lindgren, C.E., *Four-Dimensional Descriptive Geometry*, McGraw-Hill Book Company, New York, 1968.

Lindley, D., *The End of Physics - The Myth of a United Theory*, HarperCollins Publishers, Inc., 1993.

Lipschultz, M.M, *Differential Geometry,* Schaum's Outline Series, McGraw-Hill, New York, 1969.

Livio, M., *The Equation That Couldn't Be Solved – How Mathematical Genius Discovered the Language of Symmetry*, Simon and Schuster, New York, 2005.

Livio, M., *The Golden Ratio*, Broadway Books, New York, 2002.

Lockwood, E.H., *A Book of Curves*, Cambridge University Press, New York, 1961.

Lomberg, J., *Unified Force Theory, Dark Matter and Consciousness,* The Aenor Trust, PO Box 4706, Salem, Oregon, 2004.

Louis. W.C. and Van de Water, R.G., "The Darkest Particles," *Scientific American*, July 2020.

Lucas, C.W., "A Classical Electromagnetic Theory of Elementary Particles," *Journal of New Energy*, Vol. 6, No. 4, 2002.

Lucas, C.W. "A Classical Electromagnetic Theory of Elementary Particles Part 2, Interwinding Charge-Fibers," *The Journal of Common Sense Science*, Foundation of Science, May 2005, Vol. 8 No. 2.

Ludwig, C., *Michael Faraday - Father of Electronics*, Herald Press, Scottdale, PA, 1978.

Lugt, H.J., *Vortex Flow in Nature and Technology*, Krieger Publishing Company, Malabar, Florida, 1995.

Lykken, J. and Spiropulu, M., "Supersymmetry and the Crisis in Physics," *Scientific American*, May 2014.

M

Maldacena, J., "The Illusion of Gravity," *Scientific American*, pp. 57-63, November 2005.

Magueijo, J., *Faster Than the Speed of Light - The Story of a Scientific Speculation,* Penguin Books, New York, 2003.

Manning, J., *The Coming Energy Revolution - The Search for Free Energy*, Avery Publishing Group, Garden City Park, New York, 1996.

Manning, H.P., *The Fourth Dimension Simply Explained*, Dover Publications, Inc., New York, 1960.

Maor, E., *e: The Story of a Number*, Princeton University Press, Princeton, NJ, 1994.

Maor, E., *Trigonometric Delights*, Princeton University Press, Princeton, NJ, 1998.

Marsa, L., "Gene Therapy Gets Clinical," *Discover*, January/February 2020.

Marsden, J.E. and McCracken, M., *The Hopf Bifurcation and Its Applications*, Springler-Verlag, New York, Berlin, 1976.

Mazur, B., *Imagining Numbers (particularly the square root of minus fifteen)*, Picador, New York, 2003.

McCrea, W.H., "Arthur Stanley Eddington," *Scientific American*, June 1991.

McCutcheon, M., *The Final Theory – Rethinking Our Scientific Legacy*, Universal Publishers, Boca Raton, Florida, 2004.

McLeish, J., *The Story of Numbers*, Fawcett Columbine, New York, 1991.

Meacher, M., *Destination of the Species – The Riddle of Human Existence*, Books, Winchester, UK, Wasington USA, 2009.

Melker, A.A. and Krupina, M.A., "Designing Muni-Fullereneses and Their Relatives on Graph Basis," *Materials Physics and Mechanics 20*, 18-24 2014.

Mendelson, B., *Introduction to Topology*, Third Edition, Dover Publication, Inc., New York, 1990.

Messent, J., *Embroidery & Architecture*, B.T. Batsford Ltd., London, 1985.

Michalakis, S., "Quantum Leap," *Scientific American*, August 2020.

Millar, D., et al, The Cambridge Dictionary of Scientists, Cambridge University Press, 1996.

Miller, A.I., *137 – Jung, Pauli and the Pursuit of the Scientific Obsession*, W.W Norton & Company, Inc., New York, 2009.

Miller, A.I., *Albert Einstein's Special Theory of Relativity*, Springer-Verlag, New York, Berlin, 1997.

Milton, R., *Alternative Science - Challenging the Myths of the Scientific Establishment*, Park Street Press, Rochester, Vermont, 1996.

Mitchell, W.C., *Bye Bye Bing Bang – Hello Reality*, Cosmic Sense Books, Carson City, Nevada, 2002.

Mitsopoulos, T.D., "Similarity Between Elementary Particles and Electric Circuits," *Galilean Electrodynamics*, Vol. 12, No. 6, Nov./Dec. 2001.

Monro. K.S., *Love and Hope – A Message for the New Millennium*, Oughten House Publications, Livermore, California, 1997.

Montgomery, R., "The Three-Body Problem," *Scientific American*, August 2019.

Moore, W., *Schrodinger - Life and Thought*, Cambridge University Press, UK, 1992.

Moorhead, H., *The Meaning of Life*. Chicago Review Press, 1988.

Mortimer, S., *Techniques of Spiral Work – A Practical Guide to the Craft of Making Twists by Hand*, Linden Publishing, Fresno, California, 1995.

Moskowiltz, C., "The Neutrino Puzzle," *Scientific American*, October 2017.

Moskowiltz, C., "The Inner Lives of Neutron Stars," *Scientific American*, March 2019.

Moskowiltz, C., "Fusion Dreams" *Scientific American*, December 2020.

Moyer, M., "Is the Space Digital," *Scientific American*, February 2012.

Mugnai, D., et al, "Observation of Superluminal Behaviors in Wave Propagation," *Physical Review Letters*, Vol. 84, Number 21, May 2000.

Muktibodhananda, S., *Swara Yoga – The Tantric Science of Brain Breathing*, Yoga Publications Trust Munger, Bihar, India, 2006.

Murchie, G., *The Seven Mysteries of Life - An Exploration in Science and Philosophy*, Houghton Mifflin Company, Boston, 1978.

Musser, G. "Virtual Reality," *Scientific American*, September 2019.

N

Nadis, S., "Beyond Einstein." *Discover*, April 2015.

Nadis, S., "Ripple Effect." *Discover*, May 2019.

Nahin, P.J., *An Imaginary Tale – The Story of $\sqrt{-1}$*, Princeton University Press, Princeton, New Jersey, 1998.

Nakahara, M., *Geometry, Topology and Physics*, Second Edition, Taylor &Francis, Taylor & Francis Group, New York, London, 2003.

Nash, C. and Sen, S., *Topology and Geometry for Physicists*, IBI Global,

London, UK, 1983.

Natarajan, P., "The First Monster Black Holes," *Scientific American*, February 2018.

Netchitailo, V., (2020) "Hypersphere World-Universe Model: Basic Ideas." *Journal of High Energy Physics, Gravitation and Cosmology*, **6**, 710-752. doi: 10.4236/jhepgc.2020.64049.

Nernst, W., Verh. Dtsch. Phys. Ges., 18, 83, 1916.

Newton, I., *The Principia*, Prometheus Books, Amherst, New York, 1995.

Nierengarten, J-F., Editor, *Fullerenes and Other Carbon-Rich Nanostructures (Structure and Bonding)*, Springer, 2014.

Niven, W.D., *The Scientific Papers of James Clerk Maxwell*, Dover Publications, Inc., New York, 1890.

Novikov, I.D., *The River of Time*, Cambridge University Press, Cambridge, UK, 1998.

O

O'Connor, C. and Weatherall, J.O., "Why We Trust Lies," *Scientific American*, September 2019.

Okun, L.B., "The Concept of Mass," *Physics Today*, Vol. 42, June 1989.

Oliwensrein, L., "Bent out of Shape," *Discover*, July 1993.

Ornes, S., "Absolute Zero," *Discover*, July/August 2019.

Ornes, S., "The Rules of the Road to Quantum Supremacy," *Discover*, January/February 2020.

Ornes, S., "A Quantum Jump Caught in Slo-Mo," *Discover*, January/February 2020.

Oros di Bartini, R., "Relations Between Physical Constants," *Progress in Physics*, v. 3, pp. 34-40, October 2005.

Oschman, J.L. and Schman N.H., "Vortical Structure of Light and Space: Biological Implications," *J Vortex Sci Technol.* 2:1, 2015.

Oschman, J.L. and Schman N.H., "The Heart as a Bi-Directional Scalar Field Antena," *J Vortex Sci Technol.* 2:2, 2015.

Oschman, J.L., *Energy Medicine – The Scientific Basis,* Second Edition, Elsevier, New York, 2016.

P

Pagels, H.R., *The Cosmic Code – Quantum Physics as the Language of Nature*, Bantam Books, New York, 1982.

Panek, R., *The 4% Universe – Dark Matter, Dark Energy, and the Race to Discover the Rest of Reality*, Houghton Mifflin Harcourt, Boston, New York, 2011.

Panek, R., *The Trouble with Gravity: Solving the Mystery Beneath Our Feet*, Houghton Mifflin Harcourt, Boston, New York, 2019.

Panek, R., "A Cosmic Crisis," *Scientific American*, March, 2020.

Pappas, T., *The Joy of Mathematics – Discovering Mathematics All Around You,* Wide World Publishing, Tetra, 1989.

Parks, J., "Back to the Moon," *Discover*, June 2019.

Parry, A., *The Russian Scientist*, The Macmillan Company, New York, 1973.

Parson, A.L., "A Magneton Theory of the Structure of the Atom," Smithsonian Miscellaneous Collections, Vol. 65, No. 11, Publication 2371, Nov. 29, 1915.

Pauli, W., *Theory of Relativity*, Pergamon Press, London, UK, 1958.

Peat, F.D., *Superstrings and the Search for The Theory of Everything*, Contemporary Books, Lincolnwood (Chicago), ILL,1988.

Pedoe, D., *Geometry - A Comprehensive Course*, Dover Publications, Inc., New York, 1988.

Peebles, P.J.E., *Principles of Physical Cosmology*, Princeton University Press, Princeton, New Jersey, 1993.

Peirce. P,. *Frequency – The Power of Personal Vibration,* Atria Books, New York, 2009.

Penrose, R., "Twistor Quantization and Curved Space-time." *International Journal of Theoretical*

Physics (Springer Netherlands), **1**: 61-99, 1968.

Penrose, R., *The Road to Reality - A Complete Guide to the Laws of the Universe*, Alfred A. Knopf, New York, 2005.

Penrose, R., *Shadows of the Mind – A Search for the Missing Science of Consciousness,* Oxford University Press, 1994.

Peratt, A.L., "Birkeland and the Electromagnetic Cosmology," *Sky & Telescope*, May 1985.

Physical Review D: Particles and Fields, Vol. 54, The American Physical Society, 1996.

Petters, A.O., Levine, H. and Wambsganss, J., *Singularity Theory and Gravitational Lensing*, Birkhauser, Boston, Basel, Berlin, 2001.

Pierson, H.O., *Handbook of Carbon, Graphite, Diamond and Fullerenes: Properties and Applications (Material Science and Process Technology)*, Noyes Publications, 1994.

Pickover, C.A., *Mathematics and Beauty II; Spirals and "Strange" Spirals in Civilization, Nature, Science, and Art*, IBM Thomas J. Watson Research Center, Yorktown Heights, NY, 1987.

Pickover, C.A., *A Beginner's Guide to Immortality – Extraordinary People, Alien Brains, and Quantum Resurrection,* Thunder's Mouth Press, NY, 2007.

Pinkel. B,, *Consciousness. Matter, and Energy – The Emergence of Mind and Nature,* Turover Press, Santa Monica, CA. 1992.

Pollack, G. and Pollack, E, *The Fourth Phase of Water: Beyond Solid, Liquid and Vapor,* Ebner & Sons Publishers, Seattle WA, USA, 2013.

Polyakov, A., "Gauge Fields and Strings," Harwood Academic Publishers 1987, *Nucl. Phys.* **B396**, 367, 1993.

Ponomarev, C.D. and Andreeva, L.E., *The Calculation of Elastic Elements of Machines and Sensors,* Machinostroenie, Moscow, 1980.

Pontzer, H., "The particle code," *Scientific American*, January 2019.

Popkin, G., "Defying Gravity," *Discover*, April 2015.

Porter, R., *The Biographical Dictionary of Scientists*, Oxford University Press, New York, 1994.

Posamentier, A.S. and Lehmann, I., *A Biography of the World's Most Mysterious Number*, Prometheus Books, Amherst, New York, 2004.

Potemra, T. A., "Hannes Alfven, Father of Space Plasma Physics," Geomagnetism and Aeronomy with Special Historical Case Studies, IAGA Newsletters 29/1997, Published by IAGA, Germany, p.101, 1997.

Powell, D., "Einstein's Eclipse," *Discover*, May 2019.

Powell, D., "It's All Relative," *Discover*, May 2019.

Powis, R.L., *The Human Body and Why It Works,* Prentice-Hall, Inc., Englewood Cliffs, NJ, 1985.

Prasolov, V.V., *Intuitive Topology*, American Mathematical Society, USA, 2011.

Price, W.C., et al, *Wave Mechanics; The First Fifty Years - A Tribute to Professor Louis De Broglie*, John Wiley & Sons, New York-Toronto, 1973.

Price, H., *Time's Arrow and Archimedes' Point*, Oxford University Press, New York, 1996.

Purce, J., *The Mystical Spiral- Journey of the Soul,* Thames and Hudson, 1974.

Purdy, S., and Sandak, C.R., *Ancient Greece*, Franklin Watts, New York, 1982.

Puthoff, H.E., et al, "Engineering the Zero-Point Field and Polarizable Vacuum for Interstellar Flight," *Journal of New Energy*, Vol. 6, No. 1, 2001.

R

Raichlen, D.A. and Alexander, G.E., "Why Your Brain Needs Exercise," *Scientific American*, January 2020.

Randless, J. *Breaking the Time Barrier*, Paraview Pocket Books, New York, London, 2005,

Rebigsol, C.Y., *Mathematical Invalidity of Relativity (both Special and General,* 1996.

Redd, N.T. "Protecting Mars from…Ourselves" *Discover*, June 2019.

Redd, N.T. "Where Did Earth's Water Come From?" *Discover*, November 2019.

Reed, D., "Excitation and Extraction of Vacuum Energy Via EM-Torsional Field Coupling - Theoretical Model," *Journal of New Energy*, Vol. 3, No. 2/3, 1998.

Reed, D., "A New Paradigm for Time – Evidence from Empirical and Esoteric Sources," *Journal of New Energy*, Vol. 6, No. 2, 2001.

Rees, M., "Our Place in the Universe," *Scientific American*, August 2020.

Resnick, R., *Introduction to Special Relativity*, Jon Wiley & Sons, Inc., New York, London, 1968.

Ridley, B.K., *Time, Space and Things*, Cambridge University Press, Cambridge, UK, 1994.

Riordan, M., *The Hunting of the Quark - A True Story of Modern Physics*, Simon and Schuster/Touchstone, New York, 1987.

Riordan, M. and Schramm, D.N., *The Shadows of Creation - Dark Matter and the Structure of the Universe*, W.H. Freeman and Company, New York, 1991.

Ross, H., *Navigating Genesis – A Scientific Guide to Genesis 1-11*, rtb Press, 2014,

Roth, G., *The Matrix Repatterning Program for Pain Relief – Self-Treatment for Musculoskeletal Pain. New, Clinically Proven Solutions*, Ner Harbinger Publications, Inc., Oakland California. 2005.

Rovelli, C., *Reality is Not What It Seems – The Journey to Quantum Gravity*, Riverhead Books, New York, 2017.

Rucker, R., *The Fourth Dimension*, Houghton Mifflin Company, Boston, 1984.

Russell, P., *The White Hole in Time*, Harper San Francisco, 1992.

Russell, W. *The Universal One*, University of Science and Philosophy, Swannanoa, Waynesboro, Virginia, 1974.

Russell, W., *A New Concept of the Universe*, The University of Science and Philosophy, Swannanoa, Waynesboro, VA, 1989.

Russell, W., *The Secret of Light*, The University of Science and Philosophy, Swannanoa, Waynesboro, VA, 1994.

Ryu, C., *The Grand Unified Theory – A Scientific Theory of Everything*, PublishAmerica, Baltimore, 2004.

S

Saito, R., Author, Editor, *Physical Properties of Carbon Nanotubes*, Imperial College Press, London, 1998.

Salem, K.G., *The New Gravity - A New Force - A New Mass - A New Acceleration - Unifying Gravity with Light*, Salem Books, Johnstown, PA, 1994.

Sagan, C., *Cosmos,* Ballantine Books, New York, 1980.

Sanders, P.A. Jr., *Scientific Vortex Information*, Free Soul Publishing, Sedona, AZ, 1992.

Sano, C., "Twisting & Untwisting of Spirals of Aether and Fractal Vortices Connecting Dynamic Aethers," *Journal of New Energy*, Vol. 6, No. 2, 2001.

Sarg, S., "A Physical Model of the Electron According to the Basic Structure of Matter Hypothesis," *Physics Essays*, Vol. 16, No.2, 180-195, 2003.

Sarg, S., *Basic Structure of Matter – Supergravitational Unified Theory*, Trafford Publishing, North America & International, 2006.

Sarg, S, "Basic Structure of Matter – Supergravitation Unified Theory Based on an Alternative Concept of the Physical Vacuum," Proceedings of the 17th Annual Conference of the NPA at Long Beach, CA, Vol. 7, pp. 479-484, 23-26 June, 2010.

Satz, H., *Before Time Began – The Big Bang & the Emerging Universe,* Oxford University Press, United Kingdom, 2017.

Savov, E., *Theory of Interaction - The Simplest Explanation of Everything*, Geones Books, Sofia,

Bulgaria, 2002.

Scharping, N., "The Future of Fertility," *Discover*, May 2019.

Schapring, N., "Hurricanes," *Discover*, July/August 2019.

Scharf, C., "The Galactic Archipelago." *Scientific American*, January 2020.

Schmidt M., "EEG's Eke Out Buried Brain Activity," *Discover*, January/February 2020.

Schneider, M., *Vision for Life – Ten Steps to Natural Eyesight Improvement, North Atlantic Books,* Berkeley, CA, 2012.

Schneider, M.S., *A Beginner's Guide to Constructing the Universe*, Harper Perennial, New York, 1995.

Schweighauser, C.A., *Astronomy from A to Z – A Dictionary of Celestial Objects and Ideas,* Sangamon State University, Springfield, Illinois, 1991.

Schwenk, T., *Sensitive Chaos – The Creation of Flowing Forms in Water and Air,* Rudolf Steiner Press, Hillside House, East Sussex, 2008.

Schwerdtfeger, H., *Geometry of Complex Numbers – Circle Geometry, Moebius Transformation, Non-Euclidean Geometry*, Dover Publication, Inc., New York, 1979.

Schwartz, J., "Underwater," *Scientific American*, August 2018.

Scientific American, *The Enigma of Weather*, Scientific American, New York, NY, October 2004.

Scientific American, *Humans – Why We're Unlike Any Other Species on the Planet*, Scientific American, New York, NY, September 2018.

Scoles, S., "Pulsars," *Discover*, July/August 2019.

Segal, V.M., "Materials Processing by Simple Shear," *Mat. Sci & Eng.,* vol. 197, 157-164, 1995.

Segal, V.M., "Equal Channel Angular Extrusion: From Macro Mechanics to Structure Formation," *Mat. Sci & Eng.,* vol. 271, 322-333, 1999.

Segal, V.M., "Severe Plastic Deformation: Simple Shear versus Pure Shear," vol. 338, pp. 331-344, 2002.

Seggern, D.H. von, *CRC Handbook of Mathematical Curves and Surfaces*, CRC Press, Boca Raton, Florida, 1990.

Seggern, D.H. von, *CRC Standard Curves and Surfaces*, CRC Press, Boca Raton, Florida, 1993.

Segre, E., *Nuclei and Particles - An Introduction to Nuclear and Subnuclear Physics*, W.A. Benjamin, Inc., New York, 1965.

Seife, C., *Zero - The Biography of a Dangerous Idea*, Viking Penguin, New York, 2000.

Semat, H., *Introduction to Atomic and Nuclear Physics,* Rinehart & Company, Inc., New York, 1958.

Series, G.W., *Advances - The Spectrum of Atomic Hydrogen*, World Scientific, New Jersey, 1988.

Serway, R.A., *Physics for Scientist & Engineers with Modern Physics*, 3rd Edition, Saunders Golden Sunburst Series, Saunders College Publishing, Philadelphia, PA, 1990.

Seward, C., "Ball Lightning Events Explained as Self-Stable Spinning High-Density Plasma Toroids or Atmospheric Spheromaks," IEEE *Access* Practical Innovations, Volume 2, 2014, 153-59.

Sharlin, H.I., *Lord Kelvin - The Dynamic Victorian*, The Pennsylvania State University Press, PA, 1979.

Sheka, E., *Fullerences: Nanochemistry, Nanomagnetism, Nanomedicine, Nanophotonics,* CRC Press, 2011.

Sharkey, N., "Autonomous Warfare," *Scientific American*, February 2020.

Siegel. M., *The Inner Pulse – Unlocking the Secret Code of Sickness and Health,* John Wiley & Sons, Hoboken, New Jersey, 2011.

Siegfried, T., *Strange Matters – Undiscovered Ideas at the Frontiers of Space and Time,* The Berkley Publishing Group, A division of Penguin Group, New York, 2004.

Siegfried, T., *The Bit and the Pendulum*, John Wiley & Sons, Inc., New York, 2000.

Simhony, M., *Invitation to the Natural Physics of Matter, Space, Radiation*, World Scientific, New Jersey, London, 1994.

Simulik, V., *What is the Electron?* Editor, Aperon, Montreal, Canada, Published by C. Roy Keys, Inc,

2009.

Smith, K.N., How to Build Aliens in the Lab." *Discover*, July/August 2019.

Smolin, L., *The Trouble With Physics: The Rise of String Theory, The Fall of a Science, and What Comes Next,* Houghton Mifflin, 2006.

Smolin, L., *Einstein's Unfinished Revolution: The Search for What Lies Beyond the Quantum,* Penguin Press. New York, 2019.

Sinha, U., "The Triple-Slit Experiment," *Scientific American*, January 2020.

Sprott, J.C., *Strange Attractions - Creating Patterns in Chaos*, M&T Books, New York, 1993.

Sproull, R.L., *Modern Physics – A Textbook for Engineers,* John Wiley & Sons, New York, 1956.

Stenger, V.J., *God and the Atom – From Democritus to the Higgs Boson: The Story of a Triumphant Idea*, Prometheus Books, New York, 2013.

Stern, A., "Pluto Revealed," *Scientific American*, December 2017.

Stephen, K., "Is Confrontation Inevitable?" *Scientific American*, August 2019.

Sternberg, S., *Curvature in Mathematics and Physics*, Dover Publications, Inc., New York, 2012.

Sternglass, E.J., *Before the Big Bang - The Origins of the Universe*, Four Walls Eight Windows, New York, NY, 1997.

Strogatz, S., *Sync - The Emerging Science of Spontaneous Order*, Hyperion, New York, 2003.

Sunden, O., "Time-Space-Oscillation: Hidden Mechanism Behind Physics," *Galilean Electrodynamics*, Vol. 12, Special Issue 2, Fall 2001.

Swedenborg, E., *The Principia*, Swedenborg Society, London, 1912.

Synge, J.L., "The Electrodynamic Double Helix," In *Magic Without Magic: John Archibald Wheeler* - A Collection of Essays in Honor of His Sixtieth Birthday, edited by John R. Klauder, W. H. and Company, San Francisco, 1972.

T

Tanaka, K., Editor, Iijima, S., *Carbon Nanotubes and Graphene, Second Edition,* Nanotube Research Center, Tsukuba, Japan, 2014.

Talbot, M., *The Holographic Universe*, Harper Perennial, 1992.

Talcott, R., "Go Big." *Discover*, November 2019.

Tarlach, G., "Plate Tectonics," *Discover*, July/August 2019.

Tegmark, M., *Life 3.0 – Our Mathematical Universe – My Quest for the Ultimate Nature of Reality*, Vintage Books, New York, 2014.

Tegmark, M., *Life 3.0 – Being Human in the Age of Artificial Intelligence*, Vintage Books, New York, 2017.

Temes, P.S. and Rotar, F., *We the People – Human Purpose in Digital Age*, Insight Editions, New York, 2019

Tewari, P., *Universal Principles of Spacetime and Matter - A Call for Conceptual Reorientation,* Crest Publishing House, New Deli, 2002.

Tewari, P., "On the Space-Vortex Structure of the Electron," www.tewari.org/ Theory_Papers/Tewari-Final%20Proof.pdf. 2005.

Thomson, J.J., *A Treatise on the Motion of Vortex Rings*, MacMillan and Co., London, 1883.

Thomson, J.J., *Electricity and Matter*, Charles Scribner's Sons, New York, 1904.

Thomson, D.W. and Bourassa, J.D., *Secrets of Aether*, Published by The Aenor Trust, Salem, OR, 2004.

Time-Life Books, *A Soaring Spirit - Time Frame BC 600-400*, The Time Inc. Book Company, Alexandria, VA, 1987.

Time-Life Books, *Empires Ascendant - Time Frame 600 BC - AD 200*, The Time Inc. Book Company, Alexandria, VA, 1987.

Tricker, R.A., *The Contributions of Faraday and Maxwell to Electrical Science*, Pergamon Press Ltd.,

London, UK, New York, 1987.

Thorne, K.S., *Black Holes & Time Warps - Einstein's Outrageous Legacy*, W.W. Norton & Company, New York, 1994.

Tolle E., *A New Earth – Awakening to Your Life's Purpose*, Penguin's Books, 2005.

Tullis, P., "GPS Down," *Scientific American*, December 2019.

Treasures of Early Irish Art: 1500 B.C. to 1500 A.D., From the Collections of the National Museum of Ireland, Royal Irish Academy, Trinity College, Dublin, 1977.

U

Unger, R.M. and Smolin, L., *The Singular Universe and the Reality of Time*, Cambridge University Press, Cambridge, UK, 2015.

Unzicker, A., *The Higgs Fake: How Particle Physicists Fooled the Nobel Committee*, 2013.

Unzicker, A. and Jones, S., *Bankrupting Physics: How Today's Top Scientists are Gambling Away Their Credibility*, PALGRAVE MACMILLAN, New York, 2013.

V

Von Stade, S., "Owner". *Flowtoys.* Flowtoys (Toroflux).

Valens, E.G., *The Attractive Universe: Gravity and Shape of Space*, Motion, Magnet, 1969.

Van der Laan, C., "The Vortex Theory of Atoms," Thesis for the Master's Degree in History and Philosophy of Science, Institute for History and Foundation of Science, Utrecht University, Dec. 2012.

Van Eenwik, J.R., *Archetypes & Strange Attractors - The Chaotic World of Symbols*, Inner City Books, Toronto, Canada, 1997.

Van Flandern, T., *Dark Matter, Missing Planets & New Comets - Paradoxes Resolved, Origins Illuminated*, North Atlantic Books, Berkeley, CA, 1993.

Valone, T., "Inside Zero Point Energy," *Journal of New Energy*, Vol. 5, No. 4, 2001.

Veltman, M., *Facts and Mysteries in Elementary Particle Physics*, World Scientific, New Jersey, 2003.

Venable, W.M., *The Interpretation of Spectra*, Reinhold Publishing Corporation, New York, 1948.

Vineard, M., *How You Stand, How You Move, How You Live – Learning the Alexander Technique to Explore Your Mind-Body Connection and Achieve Self Mastery*, Da Capo Press, Cambridge Center. Cambridge. MA. 2007.

Vivekananda, R., *Practical Yoga Physiology – The Tantric Science of Brain Breathing*, Yoga Publications Trust Munger, Bihar, India, 2005.

Volk, G., "Toroids, Vortices, Knots, Topology and Quanta," Proceedings of the 18th Annual Conference the NPA, 6-9 at the University Maryland College Park, MD, Vol. 8, July 2011.

Vrooman, J.R., *Rene Descartes - A Biography*, G.P. Putnam's Sons, New York, 1970.

W

Wagner, O.E., "Structure in the Vacuum," *Frontier Perspectives*, Vol. 10, No. 2, Fall 2001.

Walker, F.L., "The Fluid Space Vortex: Universal Prime Mover," *Physics Essays*, Vol. 15, No. 2, 2002.

Walker, M.S., *Quantum Fuzz – The Strange True Makeup of Everything Around Us*, Prometheus Books, New York, 2017.

Wallace, D.F., *Everything and More - A Compact History of ∞*, W.W. Norton & Company, New York, London, 2003.

Walter, J., "Elon Musk Wants to Read Your Mind." *Discover*, January/February 2020.

Wardle, C., "A New World Disorder," *Scientific American*, September 2019.

Watson, J.D., *The Double Helix*, W.W. Norton & Company, New York, 1980.

Waxman, D., *The Great Life Handbook – A Practical guide to Health, Happiness, and Freedom.* Denny

Waxman Enterprises, LLC. Philadelphia, PA, 2002.

Weber, C.S., "VRML Gallery of Fullerenes," Fullerene Library, JSV1.08, 1999.

Weeks, J.R., *The Shape of Space*, 2nd ed. CRC Press, Taylor & Francis Group, Boca Raton, Florida, 2002.

Weil's A., *Breathing – The Master Key to Self-Healing,* Thorne Communications. Inc., Watertown, MA, 1999.

Weinberg, S., *Dreams of a Final Theory*, Pantheon Books, New York, 1992.

Weintraub, K., "The Means of Reproduction," *Scientific American*, March 2018.

Weir, S.T., Mitchell, A.C. and Nellis, W.J., "Metallization of Fluid Molecular Hydrogen," *Physics Review Letters* 76, 1860, 1996.

Westfall, R., *The Life of Isaac Newton*, Cambridge University Press, New York, NY, 1994.

Wheeler, J.A., *Geons, Black Holes & Quantum Foam – A Life in Physics,* W.W. Norton & Company, Inc., New York, 1998.

Wheeler, J.A., *Geometrodynamics, Topics of Modern Physics, Vol.1*, Academic Press Inc., New York, NY, 1962.

White, H.E., *Introduction to Atomic Spectra*, McGraw-Hill Book Company, Inc., New York, 1934.

Whitney, S.K., "9 Editor's Essays," *Galilean Electrodynamics*, Vol. 16, Special Issue 3, Winter 2005.

Whitney, S.K., *Algebraic Chemistry – Applications and Origins,* Nova Science Publishers, Inc., New York, 2013.

Wiener, N., *Cybernetics or Control and Communication in the Animal and the Machine*, 2nd edition, The MIT Press and John Wiley & Sons, Inc., New York, 1961.

Wigner, E. and Huntington, H.B., "On the Possibility of a Metallic Modification of Hydrogen, *Journal of Chemical Physics* **3** (12): 764, 1935.

Wilberg, K., *The Integral Vision – A Very Short Introduction to the Revolutionary Integral Approach to Life, God, the Universe, and Everything,* Shambhala, Boston & London, 2007.

Wilcock, D., *The Synchronicity Key*, A Plume Book, New York, 2014.

Wilczek, F., *Longing for the Harmonies – Themes and Variations from Modern Physics*, W.W. Norton & Company, New York, 1987.

Wilczek, F., *The Lightness of Being – Mass, Ether, and the Unification of Forces*, Basic Books, New York, 2008.

Wilczek, F., *A Beautiful Question – Finding Nature's Deep Design*, Penguin Press, New York, 2015.

Wilczek, F., "Crystals in Time, *Scientific American*, New York, NY, November 2019.

Wilkes, B.J., X-ray Vision," *Scientific American*, December 2019.

Williamson, J.G, and van der Mark, M.B., "Is the electron a Photon with Toroidal Topology?" Annales de la Foundation Louis de Broglie, Volume 22, No. 2, 133, (1997).

Winn, J.N., "Shadows of Other Worlds," *Scientific American*, March 2018.

Witten, E., "Perturbative Gauge Theory as a String Theory in Twistor Space," (http://arxiv.org/abs/hep-th/0312171)" *Common Math. Phys.* 252: 189-258 (2004).

Woit, P., *Not Even Wrong: The Failure of String Theory and the Search for Unity in Physical Law*, Basic Books, 2006.

Wolff, M., "Origin of the Mysterious Instantaneous Transmission of Events in Science," *The Cosmic Light*, Vol. 4, No. 2, Spring 2002.

Wolfram, S., *A New Kind of Science*, Wolfram Media, LLC, Champaign, IL, 2002.

Wong, H.-S. P., *Carbon Nanotubes and Graphene Device,* Cambridge University Press, Cambridge, UK, 2011.

Wood, R., "The Rise of Animals," *Scientific American*, June 2019.

Woolfson, M.M., *About Stars – Their Formation, Evolutions, Locations and Compositions*, World Scientific, New Jersey, London, 2020.

Z

Zatz, H., *Before Time Began – The Big Bang & the Emerging Universe*, Oxford University Press, Oxford, United Kingdom, 2017.

Zeman, R.K. et al., *Helical/Spiral CT - A Practical Approach*, McGraw-Hill, Inc., New York, 1995.

Zombeck, M.V., *Handbook of Space Astronomy and Astrophysics*, Cambridge University Press, Cambridge, UK, 1990.

Zwikke C., The Advanced Geometry of Plane Curves and Their Applications, Dover Publications, Inc., New York, 1994.

AUTHOR'S PUBLICATIONS

A. Modeling of magnetoelastic effect and its application for designing of force and pressure transducers used in drilling rigs

1. Ginzburg, V.B., "Magnetoelastic Self-Compensating Device for Measurement of Drilling Load MKN-1." *Machinery and Oil Equipment*, Moscow (USSR), Vol. 6, pp. 23-6, 1967.

2. Ginzburg, V.B., "The Calculation of Design Parameters of Magnetoelastic Force Transducers," *Machinery and Oil Equipment*, Moscow (USSR), Vol. 10, pp. 26-8, 1967.

3. Ginzburg, V.B. and Airapetov, V.A., "Magnetoelastic Self-Compensating Pressure Measuring Device," *Machinery and Oil Equipment*, Moscow (USSR), Vol. 2, pp. 33-35, 1968.

4. Ginzburg, V.B., "Depth Magnetoelastic Transducers," *Machinery and Oil Equipment*, Moscow (USSR), Vol. 8, pp. 24-27, 1968.

5. Ginzburg, V.B. and Ginzburg, P.B. "Non-Contact Magnetoelastic Torque Transducer," *Machinery and Oil Equipment*, Moscow (USSR), Vol. 4, pp. 25-28, 1969.

6. Ginzburg, V.B., "Magnetoelastic Pressure Transducer," *Machinery and Oil Equipment*, Moscow (USSR), Vol. 6, pp. 30-33, 1969.

7. Ginzburg, V.B., *Magnetoelastic Transducers*, Energy, Moscow (USSR), 1970.

8. Ginzburg, V.B., "New Magnetoelastic Devices for Measurement of Drilling Load MKN2 and Washing Fluid Pressure MID1," *Machinery and Oil Equipment*, Moscow (USSR), Vol. 11, pp. 24-28, 1970.

9. Ginzburg, V.B., "Increasing Accuracy of Magnetoelastic Devices for Measurement of Drilling Load," *Machinery and Oil Equipment*, Moscow (USSR), Vol. 6, pp. 29-32, 1971.

10. Ginzburg, P.B. and Ginzburg, V.B., "Improvement of Maintenance and Metrological Characteristics of Magnetoelastic Torque Measuring Devices," *Automation and Telemechanization of Oil Industry*, Vol.5, pp. 18-22, 1973.

11. Ginzburg, V.B., "The Calculation of Magnetization Curves and Magnetic Hysteresis Loops for a Simplified Model of a Ferromagnetic Body," IEEE Transaction on Magnetics, Vol. MAG-12, No. 2, March 1976.

12. Ginzburg, V.B., "The Magnetoelastic Properties of a Simplified Model of a Ferromagnetic Body in Low Magnetic Field," IEEE Transaction on Magnetics, Vol. MAG-13, No. 5, March 1977.

B. Modeling of metal rolling processes and its application for designing, modernization and startup of rolling mills

1. Ginzburg, V.B., "Dynamic Characteristics of Automatic Gage Control System with Hydraulic Actuators," *Iron and Steel Engineer*, pp. 57-65, January 1984.
2. Ginzburg, V.B., "Basic Principles of Customized Computer Models for Cold and Hot Strip Mills," *Iron and Steel Engineer,* pp. 21-35, September 1985.
3. Ginzburg, V.B. and Schmiedberg W.F., "Heat Conservation Between Roughing and Finishing Mills of Hot Strip Mills," *Iron and Steel Engineer,* pp. 29-39, April 1986.
4. Ginzburg, V.B., *Reradiating Heat Shield*, U.S. Patent No. 4,595,358, Jun. 17, 1986.
5. Kelk, G.F., Ellis, R.H. and Ginzburg, V.B. "New Developments Improve Hot Strip Shape: Shapemeter-Looper and Shape Actimeter," *Iron and Steel Engineer,* pp. 48-56, August 1986.
6. Ginzburg, V.B., "Strip Profile Control with Flexible Edge Backup Rolls," *Iron and Steel Engineer,* pp. 23-33, July 1987.
7. Ginzburg, V.B., Kaplan, N.M., James, K.L. and Zickefoose, W.F., "Application of Off-Line 8. Computer Model MILLMAX at Weirton Steel's Hot Strip Mill," *Iron and Steel Engineer,* pp. 24-33, June 1988.
9. Ginzburg, V.B, *Steel-Rolling Technology: Theory and Practice*, Marcel Dekker, New York, 1989.
10. Ginzburg, V.B., Bakhtar, F. and Dittmar, R.W., "Theory and Design of Reheating Type Heat Retention Panels," *Iron and Steel Engineer,* pp. 17-25, December 1989.
11. Ginzburg, V.B., Kaplan, N.M., Bakhtar, F. and Tabone, C.J., "Width Control in Hot Strip Mills," *Iron and Steel Engineer,* pp. 25-39, June 1991.
12. Ginzburg, V.B., and Di Giusto, B., "Self-Compensating Back-up Rolls for the Improvement of Strip Profile," *Metallurgical Plants and Technology International*, pp. 98-100, Vol 2, 1993.
13. Ginzburg, V.B, *High-Quality Steel Rolling: Theory and Practice,*, Marcel Dekker, New York, 1993.
14. Ginzburg, V.B., Fanchini, R., Bakhtar, F.A., and Azzam, M., "Selection of Optimum Mill Configurations for Cold Mills," *AISE Steel Technology*, November 1999.
15. Ginzburg, V.B., *Continuous Spiral Motion System for Rolling Mills*, U.S. Patent No. 5,970,771, Oct. 26, 1999.
16. Ginzburg, V.B, and Ballas, R., *Flat Rolling Fundamentals*, Marcel Dekker, New York, 2000.
17. Ginzburg, V.B., *Superlarge Coil Handling System*, U.S. Patent No. 6,009,736, Jan. 4, 2000.
18. Ginzburg, V.B., *Continuous Spiral Motion System and Roll Bending System for Rolling Mills*, U.S. Patent No. 6,029,491, Feb. 29, 2000.
19. Ginzburg, V.B., *Metallurgical Design of Flat Rolled Steels*, Marcel Dekker, New York, 2005.
20. Ginzburg, V.B., Author and Editor, *Flat Rolled Steel Processes – Advanced Technologies*, Taylor & Francis Group LLC, Boca Raton, Florida, 2009.
21. Ginzburg, V.B., Author and Editor, *The Making, Shaping and Treating of Steels, Flat Products Volume*, Association of Iron & Steel Technology (AIST), Warrendale, PA, 2014.
22. Ginzburg, V.B. and Kaplan, N.M., Editors, *Advanced Steel Plants for Production of Hot-Rolled Flat Product,* Association of Iron & Steel Technology (AIST), Warrendale, PA, 2018.
23. Ginzburg, V.B. and Kaplan, N.M., Editors, *Advanced Equipment for Production of Hot-Rolled Flat Product,* Association of Iron & Steel Technology (AIST), Warrendale, PA, 2019.

C. Modeling of 4d spacetime structures and properties of entities of nature

1. Ginzburg, V.B., *Spiral Grain of the Universe - In Search of the Archimedes File*, University Editions, Inc., Huntington, WV, 1996.
2. Ginzburg, V.B., "Toroidal Spiral Field Theory," *Speculations in Science and Technology* 19 (3) (1996), 165-173.
3. Ginzburg, V.B., "Structure of Atoms and Fields," *Speculations in Science and Technology* 20 (1), (1997), 51-64.
4. Ginzburg, V.B., "Double Helical and Double Toroidal Spiral Fields," *Speculations in Science and Technology*, 21 (2) (1998), 79-89.
5. Ginzburg, V.B., *Unified Spiral Field and Matter - A Story of a Great Discovery*, Helicola Press, Pittsburgh, PA, 1999.
6. Ginzburg, V.B., "Nuclear Implosion," *Journal of New Energy*, Vol. 3, No. 4, 1999.
7. Ginzburg, V.B., "Dynamic Aether," *Journal of New Energy*, Vol. 6, No. 1, 2001.
8. Ginzburg, V.B., *The Unified Spiral Nature of the Quantum & Relativistic Universe*, First edition, Helicola Press, Pittsburgh, PA, 2002.
9. Ginzburg, V.B., *The Unification of Strong, Gravitational & Electric Forces*, Helicola Press, Pittsburgh, PA, 2003.
10. Ginzburg, V.B., "Electric Nature of Strong Interactions," *Journal of New Energy*, Vol. 7, No. 1, 2003.
11. Ginzburg, V.B., "Unified Spiral Field Theory – A Quiet Revolution in Physics," *VIA-Vision in Action*, Vol. 2, No. 1 & 2, 2004.
12. Ginzburg, V.B., "The Relativistic Torus and Helix as the Prime Elements of Nature," *Proceedings of the Natural Philosophy Alliance*, Vol. 1, No. 1, Spring 2004.
13. Ginzburg, V.B., *Prime Elements of Ordinary Matter, Dark Matter & Dark Energy*, Helicola Press, Pittsburgh, PA, 2006.
14. Ginzburg, V.B., *Prime Elements of Ordinary Matter, Dark Matter & Dark Energy – Beyond Standard Model & String Theory*, Second revised edition, Universal Publishers, Boca Raton, Florida, 2007.
15. Ginzburg, V.B., "The Unification of Forces," *Proceedings of the Natural Philosophy Alliance*, (Paper presented at the 14th Annual Conference of the NPA, the University of Connecticut at Storrs, Connecticut, 21-25 May 2007), 4.
16. Ginzburg, V.B., "The Origin of the Universe, Part 1: Toryces," *Proceedings of the Natural Philosophy Alliance*," (Paper presented at the 17th Annual Conference of the NPA, the California State University, Long Beach, California, 23-26 June 2010), 7.
17. Ginzburg, V.B., "Basic Concept of 3-Dimensional Spiral String Theory (3D-SST)," (Paper presented at the 18th Annual Conference of the NPA, the University of Maryland, College Park, Maryland, 6-9 July 2011), 8.
18. Ginzburg, V.B., *The Spacetime Origin of the Universe*, First edition, Helicola Press, Pittsburgh, PA, 2013.
19. Ginzburg, V.B., "A Novel Method of Modeling of Fundamental Properties of Materials," Contributed papers from *MS&T15 Materials Science & Technology*, Greater Columbus Convention Center, Columbus, Ohio, USA, October 4-8, 2015.
20. Ginzburg, V.B., *The Spacetime Origin of the Universe with Visible Dark Matter & Energy*, Third edition, Helicola Press, Pittsburgh, PA, 2016.
21. Ginzburg, V.B., "A Novel Method of Modeling of Fundamental Properties of Materials," (Paper presented at *MS&T17 Materials Science & Technology Conference*, Lawrence L. Convention Center, Pittsburgh, Pennsylvania, USA, October 8-12, 2017).

22. Ginzburg, V.B., "The Reverse Direction of Modeling of Material Properties," (Paper presented at *MS&T18 Materials Science & Technology Conference*, Columbus, Ohio, USA, October 14-18, 2018).

23. Ginzburg, V.B., *The Unique Properties of 4D Spiral Spacetime Toryx,* First edition, Helicola Press, Pittsburgh, PA, 2018.

24. Ginzburg, V.B., *The 4D Spiral Spacetimes Toryx & Helyx – Prime Elements of the Multiverse*, First edition, Helicola Press, Pittsburgh, PA, 2018.

25. Ginzburg, V.B., *Method and Apparatus for Reverse Engineering of Material Properties,* Provisional Application for Patent, Application Number 62/764,099, Filing date 07/19/2018, Confirmation NO. 3331.

26. Ginzburg, V.B., *Method and Apparatus for Making Materials with a Wide Range of Physical Properties and Capable of Emitting and Absorbing the Superluminal Rays,* Provisional Application for Patent, Application Number 62/973,749, Filing date 10/24/2019, Confirmation NO. 3843.